PREFACE

A land mark in value distribution theory of holomorphic maps is the paper [33] of Griffiths and King extending Carlson and Griffiths [8]. Outstanding new results are obtained. A defect relation for holomorphic maps of affine algebraic varieties into projective algebraic varieties is established. However, [33] is not easily understood.

During the Spring of 1974, I lectured to my students on the topic of [33] to explain and to clarify the theory. Notes [82] were written in the Spring of 1975. This is an abbreviated and condensed version of [82]. The theory has been extended to parabolic spaces whose elementary properties are studied. New results and new concepts appear, as for instance the Ricci function, the Jacobian sections and the dominator, which are fundamental to the derivation of the defect relation. In devising this structure I tried to bring out the internal beauty of the subject matter and to exhibit the close connections to the earlier theory in [2], [89] and [67].

I strived to be clear and precise in the proofs. The results of Tung [87] helped immensely. To facilitate matters a detailed index, a list of general assumptions and references have been provided. Hopefully, these notes will help to clarify the subject matter and provide new insights.

I wish to thank Madelyn King for typing this manuscript. The research was partially supported by the National Science Foundation successively under the Grants NSF GP 20139 and MPS 75-07086.

Wilhelm Stoll
University of Notre Dame

For those who want to have more details and more supporting material, [82] can be obtained from the Mathematics Department of the University of Notre Dame, Notre Dame, Indiana, 46556, USA against a charge to cover expenses. Only a limited number of copies is available.

Lecture Notes in
Mathematics

Edited by A. Dold and B. Eckmann

600

Wilhelm Stoll

Value Distribution on Parabolic Spaces

Springer-Verlag
Berlin · Heidelberg · New York 1977

Author
Wilhelm Stoll
University of Notre Dame
Department of Mathematics
P.O. Box 398
Notre Dame, Indiana 46556/USA

Library of Congress Cataloging in Publication Data

Stoll, Wilhelm.
Value distribution on parabolic spaces.

(Lecture notes in mathematics : 600)
Bibliography: p.
Includes index.
1. Analytic mappings. 2. Value distribution theory.
3. Pseudoconvex domains. I. Title. II. Title:
Parabolic spaces. III. Series: Lecture notes in
mathematics (Berlin) ; 600.
QA331.S86 515'.73 77-9874

AMS Subject Classifications (1970): 32F15, 32H25, 32H99

ISBN 3-540-08341-3 Springer-Verlag Berlin · Heidelberg · New York
ISBN 0-387-08341-3 Springer-Verlag New York · Heidelberg · Berlin

Printed in Germany
Printing and binding: Beltz Offsetdruck, Hemsbach/Bergstr.
2141/3140-543210

GERMAN LETTERS

A	B	C	D	E	F	G	H	I	J	K	L	M
𝔄	𝔅	ℭ	𝔇	𝔈	𝔉	𝔊	ℌ	ℑ	𝔍	𝔎	𝔏	𝔐
N	O	P	Q	R	S	T	U	V	W	X	Y	Z
𝔑	𝔒	𝔓	𝔔	ℜ	𝔖	𝔗	𝔘	𝔙	𝔚	𝔛	𝔜	ℨ

a	b	c	d	e	f	g	h	i	j	k	l	m
𝔞	𝔟	𝔠	𝔡	𝔢	𝔣	𝔤	𝔥	𝔦	𝔧	𝔨	𝔩	𝔪
n	o	p	q	r	s	t	u	v	w	x	y	z
𝔫	𝔬	𝔭	𝔮	𝔯	𝔰	𝔱	𝔲	𝔳	𝔴	𝔵	𝔶	𝔷

CONTENTS

I. Introduction

Let $f: M \to N$ be a holomorphic map between complex spaces. On N, a family $\mathfrak{E}$ of analytic subsets is given. Value distribution investigates the inverse family $f^{-1}(\mathfrak{E})$. Various choices of $\mathfrak{E}$ have been made. So $\mathfrak{E}$ may be the family of hyperplanes in projective space (see Ahlfors [1], Weyl [89], Stoll [67] and [80], Murray [48], Cowen-Griffiths [18], Wong [90]) or the family of p-dimensional planes in projective space (see Levine [46], Chern [10] and [11], Stoll [75]) or $\mathfrak{E}$ may be an admissible family (see Hirschfelder [35], Wu [92], Stoll [76], Tung [87]). Bott and Chern [4] let $\mathfrak{E}$ be the family of zero sets of holomorphic sections in vector bundles. Cowen [16] and Stoll [84] consider Schubert zeroes of vector bundles. Bott, Chern and Cowen take f to be the identity map. Carlson and Griffiths [8], and Griffiths and King [33] let $\mathfrak{E}$ be the family of zero sets of holomorphic sections in a holomorphic line bundle. This theory is studied here. In this introduction, a short outline of the theory will be given without mentioning of the precise assumptions. For those, the reader should consult the main body of this monograph.

Let L be a non-negative holomorphic line bundle over the complex space N. Take a hermitian metric κ along the fibers of L such that the first Chern form $c(L,\kappa)$ is non-negative. Let V be a vector space of holomorphic sections of L over N. Assume $0 < \dim V = k + 1 < \infty$. Let $\mathbb{P}(V)$ be the associated projective space and let $\mathbb{P}: V-\{0\} \to \mathbb{P}(V)$ be the projection. If $A \subseteq V$, define $\mathbb{P}(A) = \mathbb{P}(A-\{0\})$. The <u>Grassmann cone</u> of order p is defined by

$$\tilde{G}_p(V) = \{\mathfrak{a}_0 \wedge \ldots \wedge \mathfrak{a}_p \mid \mathfrak{a}_\mu \in V\} \subseteq \underset{p+1}{\Lambda} V$$

and $G_p(V) = \mathbb{P}(\tilde{G}_p(V))$ is the <u>Grassmann manifold</u> of order p. If s is any holomorphic section in a holomorphic vector bundle, let Z(s) be the zero set of s. Take $a \in G_p(V)$. Then $a \in \mathbb{P}(\mathfrak{a}_0 \wedge \ldots \wedge \mathfrak{a}_p)$ and

$$E_L(a) = Z(\mathfrak{a}_0) \cap \ldots \cap Z(\mathfrak{a}_p)$$

is well defined. The family $\mathfrak{E}_p = \{E_L[a] \mid a \in G_p(V)\}$ is studied for $p = 0,1,\ldots,k$. Here $G_k(V)$ consists of exactly one point denoted by ∞ and $E_L[\infty]$ is called the <u>base point set</u> of L. <u>Assume</u> $E_L[\infty]$ <u>is thin</u>.

Let $\eta: N \times V \to L$ be the evaluation map defined by $\eta(x,\mathfrak{a}) = \mathfrak{a}(x)$. Then η extends to

$$\eta: N \times \bigwedge_{p+1} V \longrightarrow L \otimes (N \times \Lambda_p V)$$

$$\eta(x,\mathfrak{a}_0 \wedge \dots \wedge \mathfrak{a}_p) = \sum_{\mu=0}^{p} (-1)^{\mu} \eta(x,\mathfrak{a}_\mu) \otimes (x,\mathfrak{a}^{\mu})$$

$$\mathfrak{a}^{\mu} = \mathfrak{a}_0 \wedge \dots \wedge \mathfrak{a}_{\mu-1} \wedge \mathfrak{a}_{\mu+1} \wedge \dots \wedge \mathfrak{a}_p.$$

If $\mathfrak{z} \in \Lambda_{p+1} V$, a holomorphic section $\eta_{\mathfrak{z}}$ of $L \otimes (N \times \Lambda_p V)$ is defined by $\eta_{\mathfrak{z}}(x) = \eta(x,\mathfrak{z})$. Clearly if $\mathfrak{z} \in V$, then $\eta_{\mathfrak{z}} = \mathfrak{z}$. Let ℓ be a hermitian metric on V. It induces a hermitian metric along the fiber of $N \times \Lambda_p V$. Hence a hermitian metric $\kappa \otimes \ell$ is defined along the fibers of $L \otimes (N \times \Lambda_p V)$ If $a \in G_p(V)$, take $0 \neq \mathfrak{a} \in \tilde{G}_p(V)$ such that $\mathbb{P}(\mathfrak{a}) = a$ and define

$$0 \leqq \|a,x\|_{\kappa} = |\mathfrak{a}|^{-1} |\eta_{\mathfrak{a}}(x)|_{\kappa\otimes\ell}.$$

Then $E_L[a] = \{x \in N \mid \|a,x\|_{\kappa} = 0\}$. Define $N_\infty = N - E_L[\infty]$. Then $\eta: N_\infty \times V \to L|N_\infty$ is surjective. Hence ℓ induces a hermitian metric ℓ along the fibers of $N_\infty \times V$ and $L|N_\infty$. So $0 \leqq \|a,x\|_{\ell} \leqq 1$ is defined for $x \in N_\infty$ with

$$\|a,x\|_{\kappa} = \|a,x\|_{\ell} \|\infty,x\|_{\kappa}.$$

Here $\|\infty,x\|_{\kappa}$ is called the <u>deviation</u>. If $\|\infty,x\|_{\kappa} \leqq 1$ for all $x \in N$, then κ is <u>distinguished</u> and $\|a,x\|_{\kappa} \leqq 1$ for all $a \in G_p(V)$ and $x \in N$ follows. <u>Assume κ is distinguished</u>. If N is compact, this is easily accomplished by multiplying κ with a suitable positive constant $\gamma > 0$ which does not change the Chern form. The form

$$\Phi_L[a] = c(L,\kappa) + dd^c \log \|a,\square\|_{\kappa}^2$$

does not depend on κ. (Here $\square$ is a place holder.) The <u>Chern-Levine</u>

<u>form</u> of L for $a \in G_p(V)$ is defined by

$$0 \leqq \Lambda_L[a]_\kappa = -\log \|\square a\|_\kappa^2 \sum_{\mu=0}^{p} \Phi_L[a]^\mu \wedge c(L,\kappa)^{p-\mu}.$$

Now $\Phi_L[a]^{p+1} \equiv 0$ implies $dd^c \Lambda_L[a]_\kappa = c(L,\kappa)^{p+1}$.

Let M be a complex space of pure dimension $m > 0$. The First Main Theorem is proved for condensors on M and for parabolic exhaustions. However in this introduction, the First Main Theorem shall be formulated in the case of a parabolic exhaustion only. Let $\tau \geqq 0$ be a function of class C^∞ on M. For $r \geqq s \geqq 0$ and $A \subseteq M$, define

$$A[r] = \{x \in A \mid \tau(x) \leqq r^2\} \qquad A(r) = \{x \in A \mid \tau(x) < r^2\}$$

$$A\langle r\rangle = \{x \in A \mid \tau(x) = r^2\} \qquad A(s,r] = \{x \in A \mid s^2 < \tau(x) \leqq r^2\}.$$

$$\upsilon = dd^c\tau \qquad \rho_p = d^c\tau \wedge \upsilon^p \qquad \rho = \rho_{m-1}$$

$$\omega = dd^c \log \tau \qquad \sigma_p = d^c \log \tau \wedge \omega^p \qquad \sigma = \sigma_{m-1}.$$

Then τ is said to be a <u>parabolic exhaustion</u> of M if $M[r]$ is compact for all $r > 0$, if $M[0]$ has measure zero, if $\omega \geqq 0$ (hence $\upsilon \geqq 0$) and if $\upsilon^m \not\equiv 0 \equiv \omega^m$. Then (M,τ) called a <u>parabolic space</u>. Then

$$\varsigma = \int_{M\langle r\rangle} \sigma > 0 \quad \text{and} \quad \int_{M[r]} \upsilon^m = \int_{M(r)} \upsilon^m = \varsigma\, r^{2m}.$$

For example, $(\mathbb{C}^m,\tau)$ with $\tau(\mathfrak{z}) = |\mathfrak{z}|^2$ is parabolic. If (M,τ) is parabolic, if $\tilde{M}$ is a pure m-dimensional complex space and if $\beta: \tilde{M} \to M$ is a proper, holomorphic map of strict rank m, then $(\tilde{M},\tau\circ\beta)$ is parabolic. In particular all affine algebraic spaces of pure dimensions are parabolic. If (M_j,τ_j) are parabolic, and if $\pi_j: M_1 \times M_2 \to M_j$ are the projections, $(M_1 \times M_2, \tau_1 \circ \pi_1 + \tau_2 \circ \pi_2)$ is parabolic. A parabolic space is not compact. A non-compact Riemann surface is parabolic, if and only if M is in the class $\mathfrak{O}_G$, i.e. each subharmonic function bounded above is constant. Hence the concept is consistent with the classical notation. $(\mathbb{C} - \mathbb{Z}) \times \mathbb{C}^{m-1}$ is parabolic, but not affine algebraic.

A function $\nu: M \to \mathbb{Z}$ is said to be an analytic precycle of dimension p if there exist analytic sets A and A* such that $A \subseteq \text{supp}\ \nu \subseteq A \cup A^*$

with dim $A^* < p$ and where A is empty or pure p-dimensional, moreover $\nu|\mathfrak{R}(A)$ is required to be locally constant where $\mathfrak{R}(A)$ is the set of simple points of A. For $t > 0$ the <u>counting function</u> n_ν is defined by

$$n_\nu(t) = t^{-2p} \int_{A[t]} \nu\, \upsilon^p$$

For $0 < s < r$, the valence function of ν is defined by

$$N_\nu(r,s) = \int_s^r n_\nu(t)\frac{dt}{t} .$$

Let $f\colon M \to N$ be a holomorphic map. Take an integer $p \geqq 0$ with $q = m - p \geqq 0$. For $t > 0$ the <u>spherical image</u> of f of order p for (L,κ) is defined by

$$A_p(t) = A_p(t,L,\kappa,f) = t^{-2q} \int_{M[t]} f^*(c(L,\kappa)^p) \wedge \upsilon^q .$$

The function $A_p \geqq 0$ increases. For $0 < s < r$, the <u>characteristic function</u> of f of order p for (L,κ) is defined by

$$T_p(r,s) = T_p(r,s,L,\kappa,f) = \int_s^r A_p(t)\frac{dt}{t} .$$

If $p = 1$, write $A(t) = A_1(t)$ and $T(r,s) = T_1(r,s)$.

Take $a \in G_{p-1}(V)$. The map f is said to be <u>adapted</u> to a for L if $f^{-1}(E_L[a])$ is either empty or pure q-dimensional. <u>Assume f is adapted to a for L.</u> Then an analytic precycle $\nu = \theta_p^a[L]$ of dimension q with support $f^{-1}(E_L[a])$ is associated. Define the <u>counting function</u> of f to a for L by $n_a(t) = n_a(t,L,f) = n_\nu(t)$ and the <u>valence function</u> of f to a for L by $N_a(r,s) = N_a(r,s,L,f) = N_\nu(r,s)$. If $0 < s < r$, the <u>deficit</u> of f to a for L is defined by

$$0 \leqq D_a(r,s) = D_a(r,s,L,\kappa,f) = \tfrac{1}{2} \int_{M(s,r]} f^*(\Lambda_L[a]_\kappa) \wedge \omega^{q+1} .$$

For almost all $r > 0$, the <u>compensation function</u> of f to a for L is defined by

$$0 \leqq m_a(r) = m_a(r,L,\kappa,f) = \tfrac{1}{2} \int_{M\langle r\rangle} f^*(\Lambda_L[a]_\kappa) \wedge \sigma_q \geqq 0.$$

Then the First Main Theorem holds

$$T_p(r,s) = N_a(r,s) + m_a(r) - m_a(s) - D_a(r,s)$$

and extends the definition of $m_a(r)$ to all $r > 0$. If $p = 1$, then $\omega^m \equiv 0$ and $D_a(r,s) \equiv 0$. The First Main Theorem holds in its classical form

$$T(r,s) = N_a(r,s) + m_a(r) - m_a(s).$$

The hermitian metric ℓ on V induces a hermitian metric ℓ along the fibers of $L|N_\infty$ which can be viewed as a hermitian metric along the fibers of L singular over $E_L[\infty]$. The value distribution functions exist and the First Main Theorem holds if κ is replaced by ℓ. Denote these modified functions with brackets, for instance $A[r] = A(r,L,\ell,f)$. If $p \geqq 1$ and if

$$\text{(I.1)} \qquad \frac{A_{p-1}[r]}{T_p[r,s]} \longrightarrow 0 \qquad \text{for } r \to \infty$$

then $f(M) \cap E_L[a] \neq \emptyset$ for almost all $a \in G_p(V)$. If $p = 1$, then $A_0[r] = c_0$ is constant and condition (I.1) is satisfied. Therefore $f(M) \cap E_L[a] \neq \emptyset$ for almost all $a \in \mathbb{P}(V)$, which is a theorem of the Casorati-Weierstrass type.

These results extend to meromorphic maps. However for the Defect relation, stricter assumptions have to be made. The spaces M and N are assumed to be manifolds and N is compact. The map $f\colon M \to N$ is holomorphic. In addition the Ricci function, the Jacobian section and the dominator have to be introduced.

Let $\Omega > 0$ be a form of degree 2m and class C^∞ on M. The Ricci form Ric Ω of bidegree (1,1) and class C^∞ is assigned in the usual manner. For $0 < s < r$ the Ricci function

$$\mathrm{Ric}\ (r,s,\Omega) = \int_s^r t^{1-2m} \int_{M[t]} \mathrm{Ric}\ \Omega \wedge \upsilon^{m-1} dt$$

is defined. A function $v \geqq 0$ of class C^∞ on M is defined by $\upsilon^m = v\Omega$ For almost all $0 < s < r$, the Ricci function of τ is defined by

$$\mathrm{Ric}_\tau(r,s) = \tfrac{1}{2} \int_{M\langle r\rangle} \log v\sigma - \tfrac{1}{2} \int_{M\langle s\rangle} \log v\sigma + \mathrm{Ric}(r,s,\Omega)$$

and does not depend on the choice of Ω. If $\upsilon^m > 0$ on M, then $\mathrm{Ric}_\tau(r,s) = \mathrm{Ric}_\tau(r,s,\upsilon^m)$. If $\beta\colon M \to \mathbb{C}^m$ is a proper, holomorphic map of strict rank m and if $\tau = |\beta|^2$, let υ be the branching divisor of β. Then

$$\mathrm{Ric}_\tau(r,s) = N_\nu(r,s).$$

If the $(m-1)$-dimensional component of $\beta(\mathrm{supp}\ \nu)$ is algebraic, then $N_\nu(r,s) = O(\log r)$ for $r \to \infty$. If $M = \mathbb{C}^m$, then $\mathrm{Ric}_\tau(r,s) \equiv 0$. If $T(r,s) \to \infty$ for $r \to \infty$, the Ricci defect

$$R_f = \limsup_{r\to\infty} \frac{\mathrm{Ric}_\tau(r,s)}{T(r,s)} \leqq \infty$$

is defined and independent of s.

Let K_M and K_N be the canonical bundles of M and N respectively. Let K_N^* be the dual of K_N and let K_{Nf}^* and K_{Nf} be the pullbacks. Then $K(f) = K_M \otimes K_{N,f}^*$ is called the Jacobian bundle of f. A global holomorphic section F of $K(f)$ is called a Jacobian section and F is said to be effective if $Z(F)$ is thin. Let F be an effective Jacobian section. Let $\nu_F \geqq 0$ be the divisor of F. If $T(r,s) \to \infty$, then

$$0 \leqq \Theta_F = \liminf_{r\to\infty} \frac{N_{\nu_F}(r,s)}{T(r,s)} \leqq \infty$$

is called the ramification defect of F.

Let $\Omega_N^n(U)$ be the vector space of holomorphic forms of degree n on the open subset U of N. Assume $\tilde{U} = f^{-1}(U) \neq \emptyset$. Then F acts as a homomorphism $F\colon \Omega_N^n(U) \to \Omega_M^m(\tilde{U})$. Let $(\cdot,\cdot)$ be the K_M-valued inner product between $K(f)$ and $K_{N,f}$. Take $\psi \in \Omega^n(U)$. Then ψ pulls back to a holomorphic section ψ_f of $K_{N,f}$ over $\tilde{U}$ not to be confused with $f^*(\psi)$. Then

$F[\psi] = (F,\psi_f) \in \Omega^m_M(U)$. If $Z[\psi] = \emptyset$, then $Z(F) \cap \tilde{U} = Z(F[\psi])$. The action extends to forms of degree 2n. Let $A^p_N(U)$ (respectively $A^{p,q}_N(U)$) be the forms of class C^∞ and degree p (respectively bidegree (p,q)) on U. Then F acts as a homomorphism $F\colon A^{2n}_N(U) \to A^{2m}_M(\tilde{U})$: For each integer $p \geqq 0$, define

$$i_p = (-1)^{p(p-1)/2}\, p!\left(\frac{i}{2\pi}\right)^p.$$

If $\varphi \in \Omega^n_N(U)$ and $\chi \in \Omega^n_N(U)$, the action of F on $A^{2n}(U)$ is determined by

$$F[i_n\varphi \wedge \overline{\chi}] = i_m F[\varphi] \wedge \overline{F[\chi]}.$$

If $0 < \psi \in A^{2n}(U)$, then $F[\psi] > 0$ on $\tilde{U} - Z(F)$ and

$$\operatorname{Ric} F[\psi] = f^*(\operatorname{Ric}\psi) \qquad \text{on } \tilde{U} - Z(F).$$

If m = n, a Jacobian section F = Df is uniquely defined by $F[\psi] = f^*(\psi)$ for all $\psi \in \Omega^n_N(U)$. If $m - n = q > 0$, if $\varphi \in \Omega^q_M(M)$, a Jacobian section is uniquely defined by $F_\varphi[\psi] = \varphi \wedge f^*(\psi)$. Also if $m < n$, Jacobian sections can be constructed explicitly.

Let $\nu_1,\ldots,\nu_q$ be non-negative divisors on N. For each $x \in N$ gather all divisors $\nu_{j_1},\ldots,\nu_{j_p}$ such that $x \in \operatorname{supp}\nu_{j_\mu}$. In some neighborhood U of x, the divisor ν_{j_μ} is the divisor of a holomorphic function f_μ. Then $\nu_1,\ldots,\nu_q$ are said to be in general position if $(df_1 \wedge \ldots \wedge df_p)(x) \neq 0$. A divisor $\nu \geqq 0$ is said to have normal crossings if $\nu = \nu_1 + \ldots + \nu_q$ and if $\nu_1,\ldots,\nu_q$ are non-negative divisors in general position.

Let $L_1,\ldots,L_q$ be holomorphic line bundles on N. Define $L = L_1 \otimes \ldots \otimes L_q$. Let κ_j be a hermitian metric along L_j for $j = 1,\ldots,q$. Define $\kappa = \kappa_1 \otimes \ldots \otimes \kappa_q$. Let V_j be a vector space of holomorphic sections of L_j with $\dim V_j > 0$. Let ℓ_j be a hermitian metric on V_j. Define $V = V_1 \otimes \ldots \otimes V_q$ and $\ell = \ell_1 \otimes \ldots \otimes \ell_q$. Take $a_j \in \mathbb{P}(V_j)$. Take $0 \neq \mathfrak{a}_j \in V_j$ with $\mathbb{P}(\mathfrak{a}_j) = a_j$. Then $a_1,\ldots,a_q$ are said to be in general position, if the divisors of $\mathfrak{a}_1,\ldots\mathfrak{a}_q$ are in general position. Define $a = a_1 \otimes \ldots \otimes a_q$ and assume that $a_1,\ldots,a_q$ are in general position. Let $\Omega > 0$ be a form of class C^∞ and degree 2n on N. Take $\lambda > 0$ and assume

$\lambda c(L,\kappa) + \mathrm{Ric}\,\Omega > 0$ on N. Then Griffiths and King [33] construct a positive form ξ_a of class C^∞ on $N - E_L[a]$ such that $\mathrm{Ric}\,\xi_a > 0$, such that $(\mathrm{Ric}\,\xi_a)^n$ is integrable over N and such that

$$(\mathrm{Ric}\,\xi_a)^n \geqq \prod_{j=1}^{q}(\delta \| a_j, \|_\kappa^2)^{\lambda-1}\xi_a$$

on $N - E_L[a]$ for some constant $\delta > 0$.

Let F be an effective Jacobian section of f. Define $M^+ = \{x \in M \mid \upsilon(x) > 0\}$ and $Z_a = \overset{+}{M} - f^{-1}(E_L[a])$. Define $\zeta_a \geqq 0$ by $F[\xi_a] = \zeta_a \upsilon^m$ on Z_a. For almost all $r > 0$ the integral

$$\mu_a(r) = \frac{1}{2}\int_{M\langle r\rangle} \log \zeta_a \sigma$$

exists. If $0 < s < r$, define

$$\mathrm{Ric}\,(r,s,\xi_a,f) = \int_s^r t^{1-2m}\int_{M[t]} f^*(\mathrm{Ric}\,\xi_a) \wedge \upsilon^{m-1} dt \geqq 0.$$

The <u>Second Main Theorem</u> (Theorem 17.1) asserts

$$\begin{aligned} &N_{\theta_F}(r,s) + \mathrm{Ric}(r,s,\xi_a,f) \\ &= \lambda \sum_{j=1}^{q} N_{a_j}(r,s,L_j,f) + \mathrm{Ric}_\tau(r,s) + \mu_a(r) - \mu_a(s) \end{aligned}$$

for almost all $0 < s < r$. See also Griffiths and King [33] Proposition 6.23. However, they require $m \geqq n$ which is not needed here. Also in [33] the fundamental role of the Jacobian section F is hidden. In fact Jacobian sections F are not consciously used in [33]. The Second Main Theorem corresponds, in a sense, to the Plücker Difference Formula of [80] Theorem 8.3 and Theorem 12.1.

The Carlson-Griffiths-King method to prove the Defect Relations requires another stringent assumption on the Jacobian section F. A Jacobian section F of f is said to be <u>dominated</u> by τ if for each $r > 0$ there exists a minimal constant $Y(r) \geqq 1$ such that

$$n\left(\frac{F[\psi^n]}{\upsilon^m}\right)^{1/n}\upsilon^m \leqq Y(r) f^*(\psi) \wedge \upsilon^{m-1}$$

on $\tilde{U} \cap M^+(r)$, whenever U is open in N with $\tilde{U} = f^{-1}(U) \neq \emptyset$ and whenever $0 \leqq \psi \in A_M^{1,1}(U)$. Then Y increases and is called the <u>dominator</u> of F. If the effective Jacobian section F is dominated by τ, then $m \geqq n =$ rank f. If M is connected, if m = n = rank f and if F = Df, then F is an effective Jacobian section dominated by τ with a constant dominator $Y \equiv m$. If M is connected, if $m \geqq n =$ rank f with $m - n = q$ and if $\varphi \in \Omega_M^q(M)$, then F_φ is dominated by Y if there exists a minimal constant $Y(r) \geqq 1$ for each $r \geqq 0$ such that

$$i_q \varphi \wedge \bar{\varphi} \leqq \left(\frac{Y(r)}{m}\right)^{m-q} \upsilon^q$$

on M(r) and φ can be chosen such that F_φ is effective. If $\beta: M \to \mathbb{C}^m$ is a proper, holomorphic map of strict rank m, if $\tau = |\beta|^2$ and if $m \geqq n =$ rank f, then φ exists such that F_φ is an effective Jacobian section dominated by τ with a constant dominator $y \equiv m$. Therefore domination exists under reasonable assumptions.

Now, assume $c(L, \kappa) > 0$ on N and let F be an effective Jacobian section of f dominated by τ with dominator Y. Then $T(r,s) \longrightarrow \infty$ for $r \to \infty$. Take a positive integer w. Let $p \geqq 0$ be the smallest integer such that $L^p \otimes K_N^w$ is non-negative. For each $\varepsilon > 0$, the <u>Third Main Theorem</u> (Theorem 18.13) holds

$$N_{\theta_F}(r,s) + \left(1 - \frac{p}{w}\right) T(r,s,L,\kappa,f)$$

$$\leqq \sum_{j=1}^{q} N_{a_j}(r,s,L_j,f) + \mathrm{Ric}_\tau(r,s)$$

$$+ c_1 \log T(r,s,L,\kappa,f) + c_2 \log Y(r) + c_3 \log r$$

for all $r > s$ with $r \in \mathbb{R}_+ - E_\varepsilon$, where $\int_{E_\varepsilon} x^\varepsilon dx < \infty$. Here c_1 and $c_2 = (1 + \varepsilon/2)\varsigma n$ and $c_3 = 2\varepsilon\varsigma n$ are constants.

Now, assume $L_1 = \dots = L_q$. This simplifies the assumptions and it may be proper to recount them here. Let N be a compact, connected, complex manifold of dimension $n > 0$. Let L be a positive holomorphic line bundle on N. Let V be a vector space of holomorphic sections of L over N. Take a hermitian metric ℓ on V. Let $\eta: N \times V \to L$ be the evaluation map. Take a distinguished hermitian metric κ along the fibers of L such that $c(L,\kappa) > 0$. Let $p \geqq 0$ be the infinium of all quotients v/w where $v \geqq 0$ and $w \geqq 1$ are integers such that $L^v \otimes K_N^w$ is non-negative. Let $a_1,\ldots,a_q$ be given in $\mathbb{P}(V)$ such that $q > p$ and such that $a_1,\ldots,a_q$ are in general position for L. Let (M,τ) be a connected, complex parabolic manifold of dimension $m > 0$. Let $f: M \to N$ be a holomorphic map with an effective Jacobian section F dominated by τ. Let Y be the dominator.

Then f is adapted to all $a \in \mathbb{P}(V)$ for L and $m \geqq n = \operatorname{rank} f$. Also $T(r,s) \to \infty$ for $r \to \infty$ and the <u>dominator defect</u> Y_F and the (Nevanlinna) <u>defect</u> $\delta_f(a,L)$ are defined by

$$Y_F = \limsup_{r\to\infty} \frac{\log y(r)}{T(r,s)}$$

$$0 \leqq \delta_f(a,L) = \liminf_{r\to\infty} \frac{m_a(r)}{T(r,s)} \leqq 1$$

$$\delta_f(a,L) = 1 - \limsup_{r\to\infty} \frac{N_a(r,s)}{T(r,s)} .$$

The <u>Defect Relation holds</u>

$$\Theta_F + \sum_{j=1}^{q} \delta_f(a_j,L) \leqq p + R_f + \varsigma\, nY_F .$$

If $R_f = \infty$ or $Y_F = \infty$, this is meaningless, but in many cases $R_f = 0$ or $Y_F = 0$ or both hold. If $R_f = Y_F = 0$, then $f(M) \cap E_L[a_j] \neq \emptyset$ for at least one index j, which is a <u>Picard-Borel</u> type theorem.

Griffiths and King [33] apply their results and methods to several problems. Some of their results are generalized here. Let $g: \mathbb{R} \to \mathbb{R}_+$ be a function. The <u>order</u> of g is defined by

$$\operatorname{Ord} g = \limsup_{r\to\infty} \frac{\log^+ g(r)}{\log r} .$$

Define Ord τ = Ord Ric $(\square, s)$. If $\nu \geqq 0$ is a divisor on M, define Ord ν = Ord $N_\nu(\square, s)$ = Ord n_ν. If f: $M \to N$ is holomorphic and if $c(L, \kappa) > 0$, then Ord f = Ord $T(\square, s)$ does not depend on s, L, κ.

A connected, compact, projective algebraic manifold N of dimension n is said to be of general type (Kodaira [43], Kobayashi-Ochiai [40]) if

$$\limsup_{k \to \infty} k^{-n} \dim \Gamma(N, K_N^k) > 0.$$

Assume N is of general type. Let (M, τ) be a connected parabolic manifold of dimension n. Let f: $M \to N$ be a holomorphic map of rank n. Then Ord f $\geqq$ Ord τ. This generalizes a Theorem of Kodaira [43] and of Griffiths-King [33].

Let N be a connected, compact, projective algebraic manifold. Let L be a positive line bundle on N such that $L \otimes K_N$ is positive. Let ν be the zero divisor of a holomorphic section of L over N. Assume ν has normal crossings. Let (M, τ) be a connected, parabolic manifold of dimension n. Let f: $M \to N$ be a holomorphic map of rank n. Then f pulls back ν to a divisor ν_f on M. Then

$$\text{Ord } f \leqq \text{Max } (\text{Ord } \nu_f, \text{Ord } \tau).$$

This generalizes a result of Griffiths and King [33].

Also the results of Griffiths and King [33] on the logarithmic derivative can be extended to parabolic spaces (Theorem 21.14) and (Theorem 21.15).

Generally speaking, the following features emerge. All proofs of the First Main Theorem rest on a Residue Theorem and Stokes Theorem. For the proof of the defect relation two methods are available, the method of Weyl and Ahlfors using associated maps and the method of Carlson-Griffiths-King using the Carlson-Griffiths form ξ_a. The Second Main Theorem (i.e. Plücker Difference Formula) requires an additional concept and choice. In the Weyl theory a covariant differentiation has to be chosen, in the Griffiths theory an effective Jacobian section has to be selected. In both theories, the Third Main Theorem and the Defect Relation can only be proved if an additional severe growth restriction is imposed on these choices of the differentiation respectively the

Jacobian section. Until now, the Weyl theory has been restricted to maps into projective space, and the Griffiths theory is restricted to the case $m \geqq n = \text{rank } f$. The removal of either of these restrictions would be highly desirable, but is no easy task.

II. Preparations

1. The Chern-Levine form on projective spaces

If A is a set, define $A^n = A \times \ldots \times A$ (n-times). If A is partially ordered, denote

$$A[r,s] = \{x \in A \mid r \leqq x \leqq s\} \qquad A(r,s] = \{x \in A \mid r < x \leqq s\}$$

$$A[r,s) = \{x \in A \mid r \leqq x < s\} \qquad A(r,s) = \{x \in A \mid r < x < s\}$$

Here r and s may not belong to A, only the inequalitities must make sense. For instance, $\mathbb{R}_+ = \mathbb{R}[0,\infty)$, $\mathbb{R}^+ = \mathbb{R}(0,\infty)$ or $\mathbb{Z}_+ = \mathbb{Z}[0,\infty)$, $\mathbb{N} = \mathbb{Z}^+ = \mathbb{Z}(0,\infty)$.

Let V be a complex vector space of dimension n + 1. Then V^* is the dual vector space, $\underset{p}{\Lambda} V$ is the exterior product, $\underset{p}{\otimes} V$ is the tensor product and $\underset{p}{\oplus} V$ is the direct sum. The Grassmann cone in $\underset{p+1}{\Lambda} V$ is defined by

$$\tilde{G}_p(V) = \{\mathfrak{v}_0 \wedge \ldots \wedge \mathfrak{v}_p \mid \mathfrak{v}_\mu \in V\}$$

with $\tilde{G}_0(V) = V$ and $\tilde{G}_n(V) \approx \mathbb{C}$. For $\mathfrak{r} \in \underset{p}{\Lambda} V$ and $\alpha \in \underset{p}{\Lambda} V^* = (\underset{p}{\Lambda} V)^*$ the inner product $(\mathfrak{r};\alpha) = \alpha(\mathfrak{r})$ is defined. Let p and q be integers with $1 \leqq q \leqq p \leqq n + 1$. For $\mathfrak{r} \in \underset{p}{\Lambda} V$ and $\alpha \in \underset{q}{\Lambda} V^*$ the interior product $\mathfrak{r} \llcorner \alpha \in \underset{p-q}{\Lambda} V$ is uniquely defined by $(\mathfrak{r} \llcorner \alpha;\beta) = (\mathfrak{r};\alpha \wedge \beta)$ for all $\beta \in \underset{p-q}{\Lambda} V^*$. If p = q, this means $\mathfrak{r} \llcorner \alpha = (\mathfrak{r};\alpha)$. Take $p \in \mathbb{Z}[0,n]$ and define $q = n + 1 - p$. If $\mathfrak{e} = (\mathfrak{e}_0,\ldots,\mathfrak{e}_n)$ is a base of V, an isomorphism $D_{\mathfrak{e}} : \underset{p}{\Lambda} V \to \underset{q}{\Lambda} V^*$ is defined by

$$(\mathfrak{r};D_{\mathfrak{e}}\mathfrak{a})\, \mathfrak{e}_0 \wedge \ldots \wedge \mathfrak{e}_n = \mathfrak{r} \wedge \mathfrak{a}$$

for all $\mathfrak{r} \in \underset{q}{\Lambda} V$ and $\mathfrak{a} \in \underset{p}{\Lambda} V$. Then $D_{\mathfrak{e}}(G_{p-1}(V)) = G_{q-1}(V^*)$.

If $0 \neq \mathfrak{x} \in V$, let $\mathbb{P}(\mathfrak{x}) = \mathbb{C}\mathfrak{x}$ be the complex line spanned by $\mathfrak{x}$. If $A \subseteq V$, define $\mathbb{P}(A) = \{\mathbb{P}(\mathfrak{x}) \mid 0 \neq \mathfrak{x} \in A\}$. Then $\mathbb{P}(V)$ is the complex projective space associated to V. A holomorphic map $\mathbb{P}: V - \{0\} \to \mathbb{P}(V)$ is defined. The same symbol $\mathbb{P}$ is used for all vector spaces. Take $p \in \mathbb{Z}[0,n]$. The Grassmann manifold $G_p(V) = \mathbb{P}(\tilde{G}_p(V))$ of order p is a connected, smooth, compact submanifold of $\mathbb{P}(\underset{p+1}{\Lambda} V)$ with dimension $d(p,n) = (n-p)(p+1)$ and degree $D(p,n)$.

Take $a \in G_p(V)$. Then $\mathfrak{a} = \mathfrak{a}_0 \wedge \ldots \wedge \mathfrak{a}_p \neq 0$ exists such that $\mathbb{P}(\mathfrak{a}) = a$. A $(p+1)$-dimensional linear subspace

$$E_V(a) = E(a) = \mathbb{C}\mathfrak{a}_0 + \ldots + \mathbb{C}\mathfrak{a}_p = \{\mathfrak{z} \in V \mid \mathfrak{z} \wedge \mathfrak{a} = 0\}$$

is associated to a, independent of the choice of $\mathfrak{a}$. The associated projective space $\ddot{E}_V(a) = \ddot{E}(a) = \mathbb{P}(E_V(a))$ is smoothly imbedded into $\mathbb{P}(V)$ and called a p-plane.

Take $a \in G_p(V^*)$. Then $\alpha = \alpha_0 \wedge \ldots \wedge \alpha_p \neq 0$ exists such that $\mathbb{P}(\alpha) = a$. A $(p+1)$-codimensional linear subspace

$$E_V[a] = E[a] = \bigcap_{\mu=0}^{p} \alpha_\mu^{-1}(0) = \{\mathfrak{z} \in V \mid \alpha \;\, \mathfrak{z} = 0\}$$

is associated to a, independent of the choice of α. The associated projective space $\ddot{E}_V[a] = \ddot{E}[a] = \mathbb{P}(E_V[a])$ is a $(n-p-1)$-plane. A biholomorphic map $\delta: G_p(V) \to G_{n-p-1}(V^*)$ is defined by $E[\delta(a)] = E(a)$. Of course, δ is also defined for V^*; then $\delta = \delta^{-1}$. The map δ is called the dualism. Observe $\mathbb{P} \circ D_{\mathfrak{e}} = \delta \circ \mathbb{P}$ for any base $\mathfrak{e}$ of V.

A positive definite hermitian form ℓ on V is called a hermitian metric on V and V with ℓ is said to be a hermitian vector space. Write also $(\mathfrak{x} \mid \mathfrak{y}) = \ell(\mathfrak{x}, \mathfrak{y})$ and let $|\mathfrak{x}| = \sqrt{(\mathfrak{x} \mid \mathfrak{x})}$ be the norm of $\mathfrak{x}$. If $A \subseteq V$ and $r \geqq 0$, denote

$$A(r) = \{\mathfrak{x} \in A \mid |\mathfrak{x}| < r\}$$

$$A[r] = \{\mathfrak{x} \in A \mid |\mathfrak{x}| \leqq r\}$$

$$A\langle r\rangle = \{\mathfrak{x} \in A \mid |\mathfrak{x}| = r\}.$$

Here $V\langle r\rangle$ is a sphere oriented such that $V\langle r\rangle$ is a boundary manifold of the open ball $V(r)$ if $r > 0$.

The hermitian metric ℓ on V induces hermitian metrics on V^*, $\otimes_p V$, $\bigwedge_p V$ and $\oplus_p V$. Take $x \in \mathbb{P}(\bigwedge_{p+1} V)$ and $y \in \mathbb{P}(\bigwedge_{q+1} V)$. Then $x = \mathbb{P}(\mathfrak{x})$ and $y = \mathbb{P}(\mathfrak{y})$, and

$$\|x:y\| = \|\mathfrak{x}:\mathfrak{y}\| = \frac{|\mathfrak{x} \wedge \mathfrak{y}|}{|\mathfrak{x}|\,|\mathfrak{y}|}$$

is well defined. If $x \in G_p(V)$, then $0 \leqq \|x:y\| \leqq 1$ and

$$\ddot{E}(x) = \{y \in \mathbb{P}(V) \mid \|x:y\| = 0\}.$$

Take $x \in \mathbb{P}(\bigwedge_{p+1} V)$ and $y \in \mathbb{P}(\bigwedge_{q+1} V^*)$ with $0 \leqq q \leqq p \leqq n$. Then $x = \mathbb{P}(\mathfrak{x})$ and $y = \mathbb{P}(\eta)$, and

$$\|x,y\| = \|\mathfrak{x},\mathfrak{y}\| = \frac{|\mathfrak{x} \llcorner \eta|}{|\mathfrak{x}|\,|\eta|}.$$

If $x \in G_p(V)$ and $y \in G_q(V^*)$, then $0 \leqq \|x,y\| \leqq 1$. If $a \in G_p(V^*)$, then

$$\ddot{E}[a] = \{x \in \mathbb{P}(V) \mid \|a,x\| = 0\}.$$

$$\|a,x\| = \|\delta(a):x\| \qquad \text{if } x \in \mathbb{P}(V).$$

Define $\tau\colon V \to \mathbb{R}_+$ by $\tau(\mathfrak{z}) = |\mathfrak{z}|^2$. The exterior derivative $d = \partial + \bar\partial$ twists to

$$d^c = \frac{i}{4\pi}(\bar\partial - \partial).$$

The following differential forms are defined:

(1.1) $\qquad \upsilon = dd^c\tau > 0 \qquad \omega = dd^c \log \tau \geqq 0$

(1.2) $\qquad \rho = d^c\tau \wedge \upsilon^n \qquad \sigma = d^c \log \tau \wedge \omega^n.$

A number of important relations hold

$$(1.3)\qquad d\rho = \upsilon^{n+1} \qquad d\sigma = \omega^{n+1} = 0 \qquad \tau^{n+1}\sigma = \rho$$

$$\tau^{p+1}\omega^p = \tau\upsilon^p - p\,d\tau \wedge d^c\tau \wedge \upsilon^{p-1}$$

$$\tau^p d\tau \wedge \omega^p = d\tau \wedge \upsilon^p$$

$$\tau^p d^c\tau \wedge \omega^p = d^c\tau \wedge \upsilon^p$$

$$\upsilon^{n+1} = (n+1)\,\frac{d\tau}{\tau} \wedge \rho = (n+1)\tau^n d\tau \wedge \sigma.$$

The identity $\omega^{n+1} \equiv 0$ is fundamental to value distribution. If $r > 0$, let $j_r\colon V\langle r\rangle \to V$ be the inclusion. Then $j_r^*(\sigma) > 0$ and $j_r^*(\upsilon) = r^2 j_r^*(\omega)$. Moreover

$$\int_{V\langle r\rangle} \sigma = 1 \qquad \int_{V(r)} \upsilon^{n+1} = r^{2n+2}.$$

The hermitian metric ℓ on V defines a Fubini-Study Kaehler metric on $\mathbb{P}(V)$, whose fundamental form $\ddot{\omega} > 0$ is uniquely defined by $\mathbb{P}^*(\ddot{\omega}) = \omega$ on $V - \{0\}$, which implies $\omega^{n+1} \equiv 0$. If M is a pure m-dimensional analytic subset of $\mathbb{P}(V)$, the degree of M is defined and computes as $\int_M \ddot{\omega}$. The induced hermitian metrics on V^* and $\overset{p+1}{\Lambda} V$ define Fubini Study metrics $\ddot{\omega}_*$ and $\ddot{\omega}_p$ on $\mathbb{P}(V^*)$ and $\mathbb{P}(\underset{p+1}{\Lambda} V)$ respectively. Hence

$$D(p,n) = \int_{G_p(V)} (\ddot{\omega}_p)^{d(p,n)}.$$

For each $a \in G_p(V)$, a real analytic form $\Phi(a)$ of bidegree (1,1) is defined on $\mathbb{P}(V) - \ddot{E}(a)$ by

$$\Phi(a) = \ddot{\omega} + dd^c \log \|a:\square\|^2$$

with $\Phi(a)^{n-p} \equiv 0$. The Chern-Levine forms $\hat{\Lambda}(a)$ and $\Lambda(a)$ of bidegree

$(n-p-1, n-p-1)$ are defined on $\mathbb{P}(V) - \ddot{E}(a)$ by

$$\hat{\Lambda}(a) = \sum_{\mu=0}^{n-p-1} \Phi(a)^{\mu} \wedge \ddot{\omega}^{n-p-1-\mu} \geqq 0$$

$$\Lambda(a) = -\log \|a:\square\|^2 \hat{\Lambda}(a) \geqq 0$$

on $\mathbb{P}(V) - \ddot{E}(a)$. Then $dd^c\Lambda(a) = \ddot{\omega}^{n-p}$. The forms were introduced by Levine [46] and Chern [10]. See also [74] and [75].

Here the forms for the dual situation are needed. Take $a \in G_p(V^*)$. A real analytic form $\Phi[a]$ of bidegree $(1,1)$ is defined on $\mathbb{P}(V) - \ddot{E}[a]$ by

$$\Phi[a] = \ddot{\omega} + dd^c \log \|a,\square\|^2 \geqq 0 \tag{1.4}$$

on $\mathbb{P}(V) - \ddot{E}[a]$. Obviously, $d\Phi[a] = 0$ and

$$\Phi[a] = \Phi(\delta(a)) \qquad\qquad \Phi[a]^{p+1} \equiv 0.$$

If $a = \mathbb{P}(\alpha)$, a linear map $u_\alpha\colon V \to {}_pV^*$ is defined by $u_\alpha(\mathfrak{x}) = \alpha \llcorner \mathfrak{x}$ for all $\mathfrak{x} \in V$. Then $E[a] = \ker u_\alpha$. The map u_α factors to a holomorphic map $u_a\colon \mathbb{P}(V) - \ddot{E}[a] \to \mathbb{P}(\underset{p}{\Lambda}V^*)$ such that $u_a \circ \mathbb{P} = \mathbb{P} \circ u_\alpha$. Then $\Phi[a] = u_a^*(\ddot{\omega}_{p-1})$. See [82] Lemma 3.3. If $\alpha = \alpha_0 \wedge \cdots \wedge \alpha_p$. Define the linear map

$$\tilde{\alpha} = (\alpha_0, \ldots, \alpha_p)\colon V \to \mathbb{C}^{p+1}.$$

Then $E[a] = \ker \tilde{\alpha}$. The map $\tilde{\alpha}$ factors to a holomorphic map $\ddot{\alpha}\colon \mathbb{P}(V) - \ddot{E}[a] \to \mathbb{P}(\mathbb{C}^{p+1}) = \mathbb{P}_p$ such that $\ddot{\alpha} \circ \mathbb{P} = \mathbb{P} \circ \tilde{\alpha}$. Then $\Phi[a] = \ddot{\alpha}^*(\ddot{\omega}_{\mathbb{P}_p})$. See [82] Lemma 3.2. Here the hermitian metric on $\mathbb{C}^{p+1}$, as on each $\mathbb{C}^m$, is given by

$$(\mathfrak{x} \mid \mathfrak{y}) = \sum_{\mu=0}^{p} x_\mu \overline{y}_\mu .$$

For $a \in G_p(V^*)$, the dual Chern-Levine forms are defined by

$$\hat{\Lambda}[a] = \sum_{\mu=0}^{p} \Phi^{\mu}[a] \wedge \ddot{\omega}^{p-\mu} \geqq 0$$

$$\Lambda[a] = -\log \|a,\square\|^2 \hat{\Lambda}[a] \geqq 0$$

on $\mathbb{P}(V) - \ddot{E}[a]$. Then

$$\hat{\Lambda}[a] = \hat{\Lambda}(\delta(a)) \qquad \Lambda[a] = \Lambda(\delta(a))$$

$$dd^c\Lambda[a] = \ddot{\omega}^{p+1}.$$

Let $h: G_p(V) \to \mathbb{C}$ be an integrable function on $G_p(V)$. The integral average of h is defined by

$$I(h) = \frac{1}{D(p,n)} \int_{G_p(V)} h(\ddot{\omega}_p)^{d(p,n)}.$$

The definition extends to vector functions. If $\chi(a)$ is a differential form on a complex manifold M, then $\chi(a)(x)$ is a vector function for each fixed x. Hence $I(\chi)(x) = I(\chi(a)(x))$ is defined, if $\chi(a)(x)$ is integrable. The integral averages of $\Phi(a)^q$, $\hat{\Lambda}(a)$ and $\Lambda(a)$ have been computed in [75] Proposition 2.8, Lemma 2.10 and Theorem 2.11 page 100. The dualism leaves the integral average invariant. Hence the dual map δ gives immediately the following result:

Theorem 1.1. Take $a \in G_p(V^*)$ and $q \in \mathbb{Z}[0,p]$. Define $s = n - p$. Then

$$I(\Phi[a]^q) = \ddot{\omega}^q$$

$$I(\log \|a;\square\|^2 \Phi[a]) = \left(-\sum_{\mu=1}^{s} \frac{1}{p-q+\mu}\right)\ddot{\omega}^q$$

$$I(\hat{\Lambda}[a]) = (p+1)\ddot{\omega}^p$$

$$I(\Lambda[a]) = \left(\sum_{\nu=0}^{p} \sum_{\mu=1}^{s} \frac{1}{\nu+\mu}\right)\ddot{\omega}^p.$$

These averaging formulas will be important. Also Chern-Levine forms for Schubert varieties on Grassmann manifolds can be defined implicitly. These integral averages can be computed thanks to a theorem of Matsushima [47]. However the exact value of the coefficients is unknown. See [84].

2. Hermitian line bundles

a) Hermitian vector bundles

Let $\pi: W \to N$ be a holomorphic vector bundle of fiber dimension k over the complex space N. For $x \in N$ and $U \subseteq N$ denote $W_x = \pi^{-1}(x)$ and $W|U = \pi^{-1}(U)$. If U is open, let $\Gamma(U;W)$ be the vector space of holomorphic sections of W over U. The zero section is denoted by $\mathfrak{o}$. If $s \in \Gamma(U;W)$ the zero set $Z(s) = \{x \in U | s(x) = \mathfrak{o}(x)\}$ is analytic. Also $v = (v_1,\ldots,v_k)$ with $v_\mu \in \Gamma(U;W)$ is said to be a holomorphic frame of W over U if $v(x) = (v_1(x),\ldots,v_k(x))$ is a base of W_x for all $x \in U$. The set of holomorphic frames is denoted by $\mathfrak{F}(U,W)$. If $f: M \to N$ is a holomorphic map, the relative product $(Y,\tilde{\pi},\tilde{f})$ of (π,f) defines the pullback vector bundle $\tilde{\pi}: Y \to M$ with $f \circ \tilde{\pi} = \pi \circ \tilde{f}$; the relative product is defined up to an isomorphism. The standard model is

$$Y = \{(x,w) \in M \times W \mid f(x) = \pi(w)\}$$

where $\tilde{\pi}: Y \to M$ and $\tilde{f}: Y \to W$ are the projections.

For functions and forms of class C^h on a complex space see Tung [87] and [80]. A function κ of class C^∞ on $W \oplus W$ is called a hermitian metric along the fibers of W if for each $x \in M$ the restriction $\kappa_x: W_x \oplus W_x \to \mathbb{C}$ defines a hermitian metric on the vector space W_x. If $w \in W_x$, define $|w|_\kappa = \sqrt{\kappa(w,w)}$. Also W together with κ is called a hermitian vector bundle. If $(Y,\tilde{\pi},\tilde{f})$ is the pullback of W under the holomorphic map $f: M \to N$, then $f^*\kappa = \kappa \circ (\tilde{f} \oplus \tilde{f})$ defines a hermitian metric along the fibers of Y. If

$$0 \longrightarrow X \xrightarrow[\iota]{} W \xrightarrow[\eta]{} Z \longrightarrow 0$$

is an exact sequence of holomorphic vector bundles, κ restricts to a hermitian metric along the fibers of X. Then

$$X^{\perp} = \bigcup_{x \in N} \{w \in W_x \mid \kappa(w, \iota(y)) = 0 \ \forall\ y \in X\}$$

is a differential vector bundle on N such that $\eta: X^{\perp} \to Z$ is a differentiable vector bundle isomorphism which carries the restriction of κ to $X^{\perp}$ over to a hermitian metric along the fibers of Z, called the quotient metric of κ. The hermitian metric κ along the fibers of W induces the dual metric κ^* along the fibers of the dual vector bundle W^* and a hermitian metric κ_p along the fibers of the exterior product $\underset{p}{\Lambda}W$. If (W, κ) are hermitian vector spaces, then hermitian metrics $\kappa_1 \oplus \ldots \oplus \kappa_p$ along the fibers of $W_1 \oplus \ldots \oplus W_p$ and $\kappa_1 \otimes \ldots \otimes \kappa_p$ along the fibers of $W_1 \otimes \ldots \otimes W_p$ are defined.

b) Line bundles

At first some remarks on complex spaces and differential forms are called for. A biholomorphic map $\alpha: U_\alpha \to U'_\alpha$ of an open subset U_α of N onto an analytic subset U'_α of an open subset G_α of $\mathbb{C}^{n_\alpha}$ is called a patch. Let $j_\alpha: U'_\alpha \to G_\alpha$ be the inclusion. The patch is called smooth if $G_\alpha = U'_\alpha$. Let $\mathfrak{P}_N$ be the set of all patches and let $\mathfrak{S}_M$ be the set of all smooth patches. If χ is a form of class C^k on N, then a form $\hat{\chi}$ of class C^k on G_α is called an extension of χ if $(j_\alpha \circ \alpha)^*(\hat{\chi}) = \chi | U_\alpha$. The form χ of bidegree (p,p) on N is said to be non-negative (resp. positive) at $a \in N$ if for every holomorphic map $\varphi: H \to N$ (resp. smooth in addition) of an open neighborhood H of $0 \in \mathbb{C}^p$ with $\varphi(0) = a$, the pullback $\varphi^*(\chi)$ is non-negative at zero (respectively positive at 0 and a is a simple point of N). If a is not a simple point of N, then χ is said to be positive at a, if χ is at least continuous at a and if there exist $\alpha \in \mathfrak{P}(M)$ with $a \in U_\alpha$ and a continuous extension $\hat{\chi}$ of χ on G_α such that $\hat{\chi}$ is positive at $\alpha(a)$. (See also III 7a)

A line bundle is a vector bundle of fiber dimension 1. Let $\pi: L \to N$ be a holomorphic line bundle with a hermitian metric κ along the fibers of L. Then there exists one and only one form $c(L, \kappa)$ of class C^∞ and bidegree (1,1) on N, called the Chern form of L for κ such that the identity

$$c(L,\kappa) = - dd^c \log |s|_\kappa^2 \qquad \text{on } U - Z(s)$$

holds for all open subsets U in M and all $s \in \Gamma(U,L)$. Then

$$dc(L,\kappa) = 0 \qquad\qquad c(L^*,\kappa^*) = - c(L,\kappa).$$

If $f\colon M \to N$ is holomorphic, and if $(\tilde{L},\tilde{\kappa})$ is the pullback of (L,κ) under f, then $f^*(c(L,\kappa)) = c(\tilde{L},\tilde{\kappa})$. If (L_μ,κ_μ) are hermitian line bundles, then

$$c(L_1 \otimes \ldots \otimes L_p,\ \kappa_1 \otimes \ldots \otimes \kappa_p) = \sum_{q=1}^{p} c(L_q,\kappa_q).$$

If κ_1,κ_2 are hermitian metrics along the fibers of L, then there exists a function u of class C^∞ such that

$$c(L,\kappa_1) = c(L,\kappa_2) + dd^c u.$$

The line bundle L is said to be <u>non-negative</u> (respectively <u>positive</u>) if there exists a hermitian metric along the fibers of L such that $c(L,\kappa) \geqq 0$ (respectively $c(L,\kappa) > 0$). Write $L \geqq 0$ (respectively $L > 0$). Let $L_1 \geqq 0$ and L_2 be holomorphic line bundles on N such that $L_1^k \otimes L_2^* \geqq 0$ for some integer $k \geqq 0$. Let $[L_2/L_1]$ be the minimum of all these integers. Let $[L_2:L_1]$ be the infinium of all quotients v/w where $v \geqq 0$ and $w \geqq 1$ are integers such that $(L_1)^v \otimes (L_2^*)^w \geqq 0$. If $L_1 > 0$ and if M is compact, then $k \in \mathbb{N}$ exists such that $L_1^k \otimes L_2^* \geqq 0$.

c) The canonical bundle

Let N be a complex manifold of pure dimension n. Let T be the <u>holomorphic tangent bundle</u> and let $\overline{T}$ be its conjugate. Then $T \oplus \overline{T}$ is the complexified differential tangent bundle. Let T^* and $\overline{T}^*$ be duals. Then T^* is the <u>holomorphic cotangent bundle</u>. Then

$$(T^* \oplus \overline{T}^*)^n = \bigoplus_{p+q=n} T^{pq}$$

$$T^{pq} = \underset{p}{\Lambda} T^* \wedge \underset{q}{\Lambda} \overline{T}^* \approx \underset{p}{\Lambda} T^* \otimes \underset{q}{\Lambda} \overline{T}^*$$

are differential vector bundles. The <u>determinant bundle</u> of M is defined by $K^* = K_N^* = \underset{n}{\Lambda} T$ and the <u>canonical bundle</u> of M is defined by $K = K_N = \underset{n}{\Lambda} T^*$. Both are holomorphic line bundles and K^* is dual to K. Observe that forms of class C^k and bidegree (n,o) are exactly the section of class C^k in K.

A positive form ψ of bidegree (n,n) and of class C^∞ on N is called a volume form on N. Let $\mathfrak{B}$ be the set of all volume forms on N and let $\mathfrak{H}$ be the set of all hermitian metrics along the fibers of K. For each $p \in \mathbb{Z}_+$ define

$$i_p = \frac{i}{2\pi}^p (-1)^{p(p-1)/2} p! \tag{2.1}$$

Then there exists one and only one bijective map $h\colon \mathfrak{B} \to \mathfrak{H}$ satisfying the following property. Take $\psi \in \mathfrak{B}$. Define $\kappa_\psi = h(\psi)$. If $U \neq \emptyset$ is open in N and if φ and χ are forms of bidegree (n,o) and class C^∞ on U, then

$$i_n \varphi \wedge \overline{\chi} = \kappa_\psi(\varphi, \chi)\psi. \tag{2.2}$$

The proof is easily obtained, see [82] Theorem 4.2. If $\psi \in \mathfrak{B}$ define the <u>Ricci form</u> of ψ by

$$\operatorname{Ric} \psi = c(K, \kappa_\psi). \tag{2.3}$$

If $\alpha \in \mathfrak{S}_N$, then $\psi = \psi_\alpha i_n d\alpha_1 \wedge \dots \wedge d\alpha_n \wedge d\overline{\alpha}_1 \wedge \dots \wedge d\overline{\alpha}_n$ on U_α and

$$\operatorname{Ric} \psi = dd^c \log \psi_\alpha$$

d) Holomorphic vector bundles on Grassmann manifolds

Let V be a complex vector space of dimension n + 1. Take

$p \in \mathbb{Z}[0,n]$. The trivial bundle $G_p(V) \times V$ contains the tautological bundle

$$S_p(V) = \{(x,\mathfrak{z}) \in G_p(V) \times V \mid \mathfrak{z} \in E(x)\} \tag{2.4}$$

as a holomorphic subbundle. The quotient bundle $Q_p(V)$ exists and the classifying sequence

$$0 \longrightarrow S_p(V) \xrightarrow{j} G_p(V) \times V \xrightarrow{\eta} Q_p(V) \longrightarrow 0 \tag{2.5}$$

is obtained. The dual bundles $Q_p(V)^*$ and $Q_p(V^*)^*$ can be defined by

$$Q_p(V)^* = \{(x,\alpha) \in G_p(V) \times V^* \mid \alpha|E(x) = 0\}$$

$$Q_p(V^*)^* = \{(x,\mathfrak{z}) \in G_p(V^*) \times V \mid \mathfrak{z} \in E[x]\}.$$

If $q = n - p - 1$, then (2.5) is the pullback of

$$0 \longrightarrow Q_q(V^*)^* \longrightarrow G_q(V^*) \times V \longrightarrow S_q(V^*)^* \longrightarrow 0 \tag{2.6}$$

under the dualism $\delta: G_p(V) \to G_q(V^*)$.

If $p = 0$, then $S_0(V)^*$ is a line bundle on $\mathbb{P}(V) = G_0(V)$, called the hyperplane section bundle. If $\alpha \in V^*$, then a holomorphic section v_α of $S_0(V)^*$ on $\mathbb{P}(V)$ is defined by $v_\alpha(x) = \alpha|E(x)$. If $\alpha \neq 0$ and $a = \mathbb{P}(\alpha)$, the section v_α is a holomorphic frame of $S_0(V)^*$ over $\mathbb{P}(V) - \ddot{E}[a]$, whose dual v_α^* is defined by

$$v_\alpha^*(\mathbb{P}(\mathfrak{r})) = \frac{\mathfrak{r}}{\alpha(\mathfrak{r})} \qquad \text{if } \mathfrak{r} \in V - E[a].$$

The canonical bundle K of $\mathbb{P}(V)$ is isomorphic to $S_0(V)^{n+1}$. Let ℓ be a hermitian metric on V. Then ℓ induces hermitian metrics ℓ along the fibers of $\mathbb{P}(V) \times V, S_0(V)$ and $S_0(V)^*$ and a Fubini-Kaehler form $\ddot{\omega} > 0$ on $\mathbb{P}(V)$. Then

$$c(S_0(V),\ell) = -\ddot{\omega}$$

$$c(S_0(V)^*,\ell) = \ddot{\omega}$$

$$\mathrm{Ric}\ \ddot{\omega}^n = -(n+1)\ddot{\omega}$$

3. Classification of semi-ample line bundles

a) Simplification

Let E be a holomorphic vector bundle of fiber dimension q over a complex space N. Let V be a complex vector space of dimension k + 1. A holomorphic vector bundle homomorphism $\eta: N \times V \to E$ is said to be a simplification of E. If $v \in V$, a constant section $\vec{v}$ is defined by $\vec{v}(x) = (x,v)$. A global holomorphic section $\eta_v = \eta \circ \vec{v} \in \Gamma(N,E)$ is defined.

Let W be a finite dimensional vector space of global holomorphic sections of E. A simplification is defined by the evaluation map $e: N \times W \to E$ where $e(x,s) = s(x)$, for all $(x,s) \in N \times W$. In a sense, this example is general. If $\eta: N \times V \to E$ is a simplification, $W = \{\eta_v | v \in V\}$ is a finite dimensional vector space of sections. A holomorphic vector bundle homomorphism $\sigma: M \times V \to M \times W$ is defined by $\sigma(x,v) = (x,\eta_v)$. Then $\eta = e \circ \sigma$. The simplification η reduces to an evaluation. However, simplifications are useful.

The simplification η is said to be ample at $x \in N$, if $\eta(\{x\} x V) = E_x$. The set N_∞ of all $x \in N$ such that η is ample at x is open. The simplification is said to be ample or an amplification if $N = N_\infty$. Also $N - N_\infty$ is analytic (see [84]). In the case of a line bundle this will be shown shortly. Then η is said to be semi-ample or a semi-amplification if $N - N_\infty$ is thin. The vector bundle E is said to be ample (respectively semi-ample) if there exists an amplification (respectively semi-amplification) of E.

Let L be a holomorphic line bundle over N. Let $\eta: N \times V \to L$ be a simplification. The pth associated simplification

$$\eta: N \times \underset{p+1}{\Lambda} V \longrightarrow L \otimes (N \times \underset{p}{\Lambda} V) \tag{3.1}$$

is defined by

$$\eta(x,\mathfrak{a}_0 \wedge \dots \wedge \mathfrak{a}_p) = \sum_{\mu=0}^{p} (-1)^{\mu}\eta(x,\mathfrak{a}_\mu) \otimes (x,\mathfrak{a}^{\mu})$$

with

$$\mathfrak{a}^{\mu} = \mathfrak{a}_0 \wedge \dots \wedge \mathfrak{a}_{\mu-1} \wedge \mathfrak{a}_{\mu+1} \wedge \dots \wedge \mathfrak{a}_p.$$

Since η is skew-symmetric in $\mathfrak{a}_0,\dots,\mathfrak{a}_p$, the vector bundle homomorphism η in (3.1) is well defined. The definition of η can also be written as

$$\eta \circ \vec{\mathfrak{a}}_0 \wedge \dots \wedge \vec{\mathfrak{a}}_p = \sum_{\mu=0}^{p} (-1)^{\mu}\eta_{\mathfrak{a}_\mu} \otimes \vec{\mathfrak{a}}^{\mu}.$$

Take $a \in G_p(V)$. Then $a = \mathbb{P}(\mathfrak{a})$ with $\mathfrak{a} = \mathfrak{a}_0 \wedge \dots \wedge \mathfrak{a}_p$. The vectors $\mathfrak{a}^{\mu} \in \underset{p}{\Lambda} V$ for $\mu = 0,\dots,p$ are linearly independent. Therefore $\eta_{\mathfrak{a}}(x) = 0$ if and only if $\eta_{\mathfrak{a}_\mu}(x) = 0$ for all $\mu = 0,\dots,p$. If $0 \neq \lambda \in \mathbb{C}$, then $\eta_{\lambda\mathfrak{a}} = \lambda\eta_{\mathfrak{a}}$. Hence the zero set

(3.2) $$E_L[a] = Z(\mathfrak{a}) = Z(\eta_{\mathfrak{a}_0}) \cap \dots \cap Z(\eta_{\mathfrak{a}_p})$$

depends on a only. Obviously $E_L[a]$ is analytic with

$$E_L[a] = \bigcap_{\mathfrak{b} \in E(a)} Z(\mathfrak{b}) = \bigcap_{b \in \ddot{E}(a)} E_L[b].$$

If $f\colon M \to N$ is a holomorphic map, the value distribution of the family

$$\{f^{-1}(E_L[a]\}_{a \in G_p(V)}$$

shall be studied.

Recall $G_k(V)$ consists of exactly one point denoted by ∞. If $\alpha_0,\ldots,\alpha_k$ is a base of V, then $\infty = \mathbb{P}(\alpha_0 \wedge \ldots \wedge \alpha_k)$. Define $a_q = \mathbb{P}(\alpha_q)$ for $q = 0,\ldots,k$. Then

$$E_L[\infty] = \bigcap_{q=0}^{k} E_L[a_q] = \bigcap_{q=0}^{k} Z(\eta_{\alpha_q})$$

$$E_L[\infty] = \bigcap_{a\in\mathbb{P}(V)} E_L[a] = \bigcap_{\alpha\in V} Z(\eta_\alpha)$$

is called the <u>base point</u> set of the simplification η, obviously

$$N_\infty = N - E_L[\infty]$$

is the largest set in N such that $\eta|N_\infty$ is ample. Especially η is semi-ample if and only if $E_L[\infty]$ is thin and η is ample if and only if $E_L[\infty] = \emptyset$.

The simplification η is said to be <u>semi-ample of order p</u> or a <u>semi-amplification of order p</u> if for each branch B of N

$$\dim B - \dim_x (E_L[\infty] \cap B) > p$$

for all $x \in E_L[\infty] \cap B$. If $p \geqq \dim_x B$, then $E_L[\infty] \cap B = \emptyset$.

<u>Lemma 3.1</u>. Let $\eta: N \times V \to L$ be a semi-amplification of order $p \geqq 1$ Take $a \in G_{p-1}(V)$. Then $E_L[\infty]$ is a thin analytic subset of $E_L[a]$.

<u>Proof</u>. Obviously $E_L[\infty] \subseteq E_L[a]$. Assume a branch C of $E_L[a]$ is contained in $E_L[\infty]$. A branch B of N exists such that $B \supseteq C$. Then $E_L[\infty] \cap B \supseteq C$. Define $m = \dim B$. Take $\alpha_q \in V$ such that $a = \mathbb{P}(\alpha_1 \wedge \ldots \wedge \alpha_p)$. Then

$$B \cap E_L[a] = B \cap Z(\eta_{\alpha_1}) \wedge \ldots \wedge Z(\eta_{\alpha_p})$$

where each $Z(\eta_{\alpha_q})$ is locally the zero set of one holomorphic function. Hence $\dim_x B \cap E_L[a] \geqq m - p$ for each $x \in B \cap E_L[a]$. Contradiction, q.e.d.

If N is irreducible, then L is semi-ample if at least one global holomorphic section $s \not\equiv 0$ exists. Therefore the assumption that L is semi-ample is really no restriction.

b) Some remarks on meromorphic maps

Soon meromorphic maps will be needed. So some remarks are needed. Let M and N be complex spaces and let S be a thin analytic subset of M. Consider a holomorphic map $f: M - S \to N$. The closed graph of f is the closure F in $M \times N$ of the graph $F_0 = \{(x, f(x)) \mid x \in M - S\}$. Let $\vec{f}: F \to N$ and $\overleftarrow{f}: F \to M$ be the projections. The map f is said to be meromorphic on M and denoted by $f: M \longrightarrow N$ if F is an analytic subset of $M \times N$ and if $\overleftarrow{f}$ is proper. See Remmert [55], Stein [65] and also [2], [70], [80]. Let A be the largest open subset of M such that f continues to a holomorphic map $f_1: A \to N$. Define $\tilde{A} = \overleftarrow{f}^{-1}(A)$. Then $I_f = M - A$ and $\tilde{I}_f = F - \tilde{A}$ are thin analytic and $\overleftarrow{f}: \tilde{A} \to A$ is biholomorphic. Here I_f is called the indeterminacy of f. If M has pure dimension m, then

$$\tilde{S}_1 = \{z \in F \mid \operatorname{rank}_z \overleftarrow{f} < m\}$$

is thin analytic. The analytic set $I_f^s = \overleftarrow{f}(\tilde{S}_1)$ is called the strong indeterminacy. Then $I_f^s \subseteq I_f$ and $\dim I_f^s \leqq m - 2$. If $x \in M$, then $x \in I_f^s$ if and only if $\overleftarrow{f}^{-1}(x)$ has positive dimension. If M is normal, then $I_f^s = I_f$.

Return to the situation, where M is any complex space. For $A \subseteq M$ and $B \subseteq N$ define

$$f(A) = \vec{f}(\overleftarrow{f}^{-1}(A)) \qquad f^{-1}(B) = \overleftarrow{f}(\vec{f}^{-1}(B)).$$

If $h: X \to N$ be a holomorphic map, then

$$P = \{(z, x) \in M \times X \mid z \in f^{-1}(h(x))\}$$

is analytic. Let $\tilde{f}: P \to X$ and $\tilde{h}: P \to M$ be the projections. Then $(P, \tilde{f}, \tilde{h})$ is said to be the standard relative product of (f, h).

Let M and S as before. Let V be a complex vector space of dimension k + 1 and let $f\colon M - S \to \mathbb{P}(V)$ be holomorphic. Let $U \neq \emptyset$ be open in M. A holomorphic map $\mathfrak{v}\colon U \to V$ is said to be a representation of f on U if $\mathfrak{v}^{-1}(0)$ is thin in U and if $f(x) = \mathbb{P}(\mathfrak{v}(x))$ for all $x \in U - S$ with $\mathfrak{v}(x) \neq 0$. If $a \in U$, then $\mathfrak{v}$ is a representation at a. The representation is said to be simple if $\mathfrak{v}(x) \neq 0$ for all $x \in U$. The representation is said to be irreducible if

$$\dim_x \mathfrak{v}^{-1}(0) \leqq (\dim_x M) - 2$$

for all $x \in \mathfrak{v}^{-1}(0)$. The map $f\colon M - S \to \mathbb{P}(V)$ is meromorphic on M if and only if a representation exists at every $a \in M$. If f is meromorphic on M, a simple representation exists at a if and only if $a \in M - I_f$. An irreducible representation exists at every non-singular point of M. A meromorphic function f is a meromorphic map into $\mathbb{P}(\mathbb{C}^2)$ where $f^{-1}(\infty)$ is thin.

c) Classification

Let $\eta\colon N \times V \to L$ be a semi-amplification. Let S be the kernel of $\eta|N_\infty$. An exact sequence

$$(3.3) \qquad 0 \longrightarrow S \longrightarrow N_\infty \times V \longrightarrow L|N_\infty \longrightarrow 0$$

is defined. Here S has fiber dimension k. If $x \in N_\infty$, one and only one $\varphi_0(x) \in \mathbb{P}(V^*)$ and $\varphi_*(x) \in G_{k-1}(V)$ exist such that $E[\varphi_0(x)] = S_x = E(\varphi_*(x))$. The maps $\varphi_*\colon N_\infty \to G_{k-1}(V)$ and $\varphi_0\colon N_\infty \to \mathbb{P}(V^*)$ are called the classification map and the dual classification map respectively. If δ is the dualism, then $\varphi_0 = \delta \circ \varphi_*$. For line bundles the dual classification map is simpler, for vector bundles the classification map is more natural.

Theorem 3.2. Let $\eta\colon N \times V \to L$ be a semi-amplification. Then the dual classification map $\varphi_0\colon N_\infty \to \mathbb{P}(V^*)$ is holomorphic and extends to a meromorphic map $\varphi\colon N \rightharpoonup \mathbb{P}(V)$. The exact sequence (3.3) is the pullback of (2.6) for q = 0 under φ_0. If $v \in \mathfrak{F}(U,L)$ is a holomorphic frame of L over the open subset $U \neq \emptyset$ of N, a representation $\mathfrak{w}\colon U \to V^*$ of φ is uniquely defined by

(3.4) $$\eta_{\mathfrak{z}}(x) = (\mathfrak{w}(x);\mathfrak{z})v(x)$$

for all $x \in U$ and $\mathfrak{z} \in V$. Then $\mathfrak{w}^{-1}(0) = U \cap E_L[\infty]$ and $\varphi_0 = \mathbb{P} \circ \mathfrak{w}$ on $U \cap N_\infty$.

Proof. Given $x \in U$, the linear map $\mathfrak{w}(x)\colon V \to \mathbb{C}$ is defined by (3.4). Obviously $\mathfrak{w}(x) = 0$, iff $\eta_{\mathfrak{z}}(x) = 0$ for all $\mathfrak{z} \in V$, i.e., iff $x \in E_L[\infty]$. Let $\mathfrak{a}_0,\ldots,\mathfrak{a}_k$ be a base of V and let $\alpha_0,\ldots,\alpha_k$ be the dual base. Then $\mathfrak{w} = \sum_{\mu=0}^{k} w_\mu \alpha_\mu$ with $w_\mu = (\mathfrak{w};\mathfrak{a}_\mu)$. Hence $\eta_{\mathfrak{a}_\mu} = w_\mu v$. Therefore w_μ is holomorphic on U. Hence $\mathfrak{w}$ is holomorphic. If $x \in U \cap N_\infty$, then

$$E[\mathbb{P}(\mathfrak{w}(x))] = \{\mathfrak{z} \in V \mid (\mathfrak{w}(x);\mathfrak{z}) = 0\}$$

$$= \{\mathfrak{z} \in V \mid \eta(x,\mathfrak{z}) = 0\} = S_x = E[\varphi_0(x)]$$

Hence $\varphi_0 = \mathbb{P} \circ \mathfrak{w}$ on $U \cap N$. So φ_0 is holomorphic on N_∞ and extends to a meromorphic map $\varphi\colon N \longrightarrow \mathbb{P}(V)$ with $\mathfrak{w}$ as a representation; q.e.d.

The map $\varphi\colon N \longrightarrow \mathbb{P}(V)$ is also called the dual classification map.

Proposition 3.3. Let $\eta\colon N \times V \to L$ be a semi-amplification. Let φ be the dual classification map of η. Take $a \in G_p(V)$. Then

(3.5) $$\varphi^{-1}(E[a]) \cap N_\infty = E_L[a] \cap N_\infty$$

(3.6) $$\varphi^{-1}(\ddot{E}[a]) \subseteq E_L[a].$$

If η is a semi-amplification of order $p + 1$, then

(3.7) $$\varphi^{-1}(\ddot{E}[a]) = E_L[a].$$

Proof. Take $x_0 \in N$. An open neighborhood U of x_0 and a holomorphic frame v of L over U exist. Determine $\mathfrak{w}$ by (3.4). Take a base $\mathfrak{a}_0,\ldots,\mathfrak{a}_k$ of V with $a = \mathbb{P}(\mathfrak{a}_0 \wedge \ldots \wedge \mathfrak{a}_p)$. Let $\alpha_0,\ldots,\alpha_k$ be the dual base. Then $\mathfrak{w} = w_0\alpha_0 + \ldots + w_k\alpha_k$ and $\eta_{\mathfrak{a}_\mu} = w_\mu v$ with

$$U \cap E_L[a] = \bigcap_{\mu=0}^{p} Z(\eta_{a_\mu}) \cap U = \bigcap_{\mu=0}^{p} w_\mu^{-1}(0).$$

Take $x \in U \cap N_\infty$. Then $x \in \varphi^{-1}(\ddot{E}[a])$ means $\mathbb{P}(\mathfrak{w}(x)) = \varphi(x) \in \ddot{E}[a]$ which means $\mathfrak{w}(x) \in E[a]$ which is equivalent to $w_\mu(x) = (\mathfrak{w}(x); a_\mu) = 0$ for $\mu = 0,\ldots,p$ which is $x \in E_L[a]$. Therefore (3.5) holds. Now $N - N_\infty = E_L[\infty] \subseteq E_L[a]$ implies (3.6). If η is semi-ample of order $p + 1$, then $E_L[a] \cap N_\infty$ is dense in $E_L[a]$ by Lemma 3.1. Since $\varphi^{-1}(\ddot{E}[a])$ is closed, (3.5) implies $\varphi^{-1}(\ddot{E}[a]) \supseteq E_L[a]$. Now, (3.7) follows; q.e.d.

Let ℓ be a hermitian metric on V. Then ℓ induces a hermitian metric along the fibers of S, $N \times V$ and $L|N_\infty$ by (3.3) and along $Q_0(V^*)^*$, $\mathbb{P}(V^*) \times V$ and $S_0(V^*)^*$ by

$$(3.8) \qquad 0 \longrightarrow Q_0(V^*)^* \longrightarrow \mathbb{P}(V^*) \times V \longrightarrow S_0(V^*)^* \longrightarrow 0.$$

The dual classification map pulls back (3.8) to (3.3) under $\varphi: N \longrightarrow \mathbb{P}(V$ Hence

$$(3.9) \qquad c(L,\ell) = \varphi^*(c(S_0(V^*)^*,\ell)) = \varphi^*(\ddot{\omega}) \geqq 0.$$

Observe that the Chern form $c(L,\ell)$ is defined on N_∞ only.

4. The Chern-Levine form for line bundles

Let $\eta: N \times V \to L$ be a semi-amplification of the holomorphic line bundle L over the complex space N. Here V is a complex vector space of dimension $k + 1$ with a hermitian metric ℓ. A hermitian metric κ along the fibers of L is given. Then η extends to

$$\eta: N \times \underset{p+1}{\Lambda} V \longrightarrow L \otimes (N \times \underset{p}{\Lambda} V).$$

The hermitian metrics ℓ and κ define hermitian metrics ℓ along the fibers of $N \times \underset{p}{\Lambda} V$ and $\kappa \otimes \ell$ along $L \otimes (N \times \underset{p}{\Lambda} V)$. The norms on V, $\underset{p}{\Lambda} V$ etc.

will usually be denoted without the subscript ℓ.

Take $a \in G_p(V)$. Take $0 \neq \mathfrak{a} \in \tilde{G}_p(V)$ with $\mathbb{P}(\mathfrak{a}) = a$. The section $\eta_{\mathfrak{a}} = \eta \circ \vec{\mathfrak{a}}$ of $L \otimes (N \times \underset{p}{\Lambda} V)$ is defined (if $p = 0$ this is L). For $x \in N$, the norm $|\eta_{\mathfrak{a}}(x)|_{\kappa\otimes\ell}$ is defined. The <u>distance</u>

$$\|a, x\|_{\kappa\otimes\ell} = \|\mathfrak{a}, x\|_{\kappa\otimes\ell} = |\mathfrak{a}|^{-1} |\eta_{\mathfrak{a}}(x)|_{\kappa\otimes\ell} \geqq 0$$

depends on a and x only since $\lambda\eta_{\mathfrak{a}} = \eta_{\lambda\mathfrak{a}}$ for all $0 \neq \lambda \in \mathbb{C}$. Abbreviate

$$\|a, x\|_{\kappa} = \|a, x\|_{\kappa\otimes\ell}.$$

Especially, $\infty \in G_k(V)$ defines the <u>deviation</u> $\|\infty, x\|_{\kappa}$. The hermitian metric κ is said to be <u>distinguished</u> if and only if $\|\infty, x\|_{\kappa} \leqq 1$ for all $x \in N$. If N is compact, a constant $\lambda > 0$ exists such that $\lambda\|\infty, x\|_{\kappa}^2 \leqq 1$. Trivially

$$\|\infty, x\|_{\lambda\kappa} = \sqrt{\lambda}\, \|\infty, x\|_{\kappa} \leqq 1$$

Hence $\lambda\kappa$ is distinguished. Observe $c(L, \lambda\kappa) = c(L, \kappa)$. Hence, on compact spaces N, the metric κ can be taken to be distinguished w.l.o.g. Observe that "distinguished" depends on η and ℓ. The use of a distinguished metric will make the compensation function non-negative.

<u>Lemma 4.1</u>. Take $a \in G_p(V)$. Then

$$E_L[a] = \{x \in N \mid \|a, x\|_{\kappa} = 0\}.$$

<u>Lemma 4.2</u>. Take $a \in G_p(V)$. Let $\mathfrak{a}_0, \ldots, \mathfrak{a}_p$ be orthonormal in V such that $a = \mathbb{P}(\mathfrak{a}_0 \wedge \ldots \wedge \mathfrak{a}_p)$. Then

$$\|a, x\|_{\kappa}^2 = \sum_{\mu=0}^{p} |\eta_{\mathfrak{a}_\mu}(x)|_{\kappa}^2 .$$

<u>Proof</u>. The vectors $\mathfrak{a}^{\mu}$ are orthonormal in $\underset{p}{\Lambda} V$. Define $\mathfrak{a} = \mathfrak{a}_0 \wedge \ldots \wedge \mathfrak{a}_p$. Then

$$\|a;x\|_\kappa^2 = |\eta_{\mathfrak{a}}(x)|_\kappa^2 =$$

$$= |\sum_{\mu=0}^{p}(-1)^\mu \eta_{\mathfrak{a}_\mu}(x) \otimes \vec{\mathfrak{a}}^\mu|_\kappa^2 = \sum_{\mu=0}^{p}|\eta_{\mathfrak{a}_\mu}(x)|_\kappa^2 \quad \text{q.e.d.}$$

Lemma 4.3. Take $a = \mathbb{P}(\mathfrak{a}) \in G_p(V)$. Let v be a holomorphic frame of L over U. Define $\mathfrak{w}: U \to V^*$ by (3.4). Take $x \in U$. Then

$$\|\infty, x\|_\kappa = |\mathfrak{w}(x)||v(x)|_\kappa$$

$$\|a, x\|_\kappa = |\mathfrak{a}|^{-1}|\mathfrak{a} \llcorner \mathfrak{w}(x)||v(x)|_\kappa.$$

Proof. Take an orthonormal base $\mathfrak{a}_0, \ldots, \mathfrak{a}_k$ of V such that $\mathfrak{a} = |\mathfrak{a}|\mathfrak{a}_0 \wedge \cdots \wedge \mathfrak{a}_p$. Let $\alpha_0, \ldots, \alpha_k$ be the dual base. Then $w_\mu v = \eta_{\mathfrak{a}_\mu}$ on U and $\mathfrak{w} = w_0\alpha_0 + \ldots + w_k\alpha_k$. Hence

$$|\mathfrak{a} \llcorner \mathfrak{w}|_\kappa^2 = |\mathfrak{a}|^2|\sum_{\mu=0}^{p}(-1)^\mu w_\mu \mathfrak{a}^\mu| = |\mathfrak{a}|^2 \sum_{\mu=0}^{p}|w_\mu|^2$$

$$\|a, x\|^2 = \sum_{\mu=0}^{p}|\eta_{\mathfrak{a}_\mu}(x)|_\kappa^2 = \sum_{\mu=0}^{p}|w_\mu(x)|^2|v(x)|_\kappa^2$$

$$= |\mathfrak{a}|^{-2}|\mathfrak{a} \llcorner \mathfrak{w}(x)|_\kappa^2|v(x)|_\kappa^2.$$

Define $\mathfrak{c} = \mathfrak{a}_0 \wedge \cdots \wedge \mathfrak{a}_k$. Then $|\mathfrak{c}| = 1$ and $\mathbb{P}(\mathfrak{c}) = \infty$. Hence

$$\|\infty, x\|_\kappa^2 = \sum_{\mu=0}^{k}|w_\mu(x)|^2|v(x)|_\kappa^2 = |\mathfrak{w}(x)|^2|v(x)|_\kappa^2 \quad \text{q.e.d.}$$

Lemma 4.4. Take $a \in G_p(V)$. Let $\varphi: N \longrightarrow \mathbb{P}(V^*)$ be the dual classification map. Then

(4.1) $$\|a, x\|_\kappa = \|a, \varphi(x)\|\|\infty, x\|_\kappa \quad \text{for } x \in N$$

$$\|a, x\|_\kappa \leqq \|\infty, x\|_\kappa \quad \text{for } x \in N.$$

If κ is distinguished, then $\|a, x\|_\kappa \leqq 1$.

Proof. Take $U, \mathfrak{w}, \mathfrak{a}, \alpha$ as in the proof of Lemma 4.3. Take $x \in U \cap N_\infty$. Then $f(x) = \mathbb{P}(\mathfrak{w}(x))$. Hence

$$\|a,x\|_\kappa = \frac{|\mathfrak{a} \llcorner \mathfrak{w}(x)|}{|\mathfrak{a}|\,|\mathfrak{w}(x)|}\, |\mathfrak{w}(x)|\, |v(x)|_\kappa$$

$$= \|a,\varphi(x)\|\,\|\infty,x\|_\kappa$$

which proves (4.1). Now (4.2) follows by continuity; q.e.d.

The hermitian metric ℓ on V induces a hermitian metric ℓ along the fibers of $L|N_\infty$. Hence $\|a,x\|_\ell$ is defined for all $x \in N_\infty$.

Lemma 4.5. Take $a \in G_p(V)$ and $x \in N_\infty$. Then $\|\infty,x\|_\ell = 1$ and $\|a,x\|_\ell = \|a,\varphi(x)\|$. Especially, ℓ is distinguished and

$$\|a,x\|_\kappa = \|a,x\|_\ell \|\infty,x\|_\kappa . \tag{4.3}$$

Proof. Take an orthonormal base $\mathfrak{a}_0,\ldots,\mathfrak{a}_k$ of V such that $\mathfrak{a}_1,\ldots,\mathfrak{a}_k$ span the kernel S_x of $\eta_x\colon V \to L_x$. Then $\eta_{\mathfrak{a}_\mu}(x) = 0$ for $\mu = 1,\ldots,k$, and L_x is isometric to $S_x = \mathbb{C}\mathfrak{a}_0$. Hence $|\eta_{\mathfrak{a}_0}(x)| = |\mathfrak{a}_0| = 1$. Therefore

$$\|\infty,x\|_\ell^2 = \sum_{\mu=0}^{k} |\eta_{\mathfrak{a}_\mu}(x)|^2 = |\eta_{\mathfrak{a}_0}(x)|^2 = 1.$$

Apply Lemma 4.4; q.e.d.

The result of Lemma 4.5 explains the name deviation. The metric ℓ is the natural one, and $\|\infty,\square\|_\kappa$ shows how for κ deviates from ℓ.

Lemma 4.6. On N_∞ we have

$$c(L,\kappa) = c(L,\ell) - dd^c \log \|\infty,\square\|_\kappa^2.$$

Proof. Let v be a holomorphic frame of L over the open subset U of N_∞. Define $\mathfrak{w}\colon U \to V^* - \{0\}$ by (3.4). Then $\varphi = \mathbb{P}\circ\mathfrak{w}$ on U. Define $\tau\colon V^* \to \mathbb{R}_+$ by $\tau(\mathfrak{z}) = |\mathfrak{z}|^2$. Then $\|\infty,\square\|_\kappa^2 = (\tau\circ\mathfrak{w})|v|_\kappa^2$ on U. Hence

$$dd^c \log \|\infty,\square\|_\kappa^2 = \mathfrak{w}^*(dd^c \log \tau) + dd^c \log |v|_\kappa^2$$

$$= \mathfrak{w}^*\mathbb{P}^*(\ddot{\omega}) - c(L,\kappa)$$

$$= \varphi^*(\ddot{\omega}) - c(L,\kappa)$$

$$= c(L,\ell) - c(L,\kappa) \qquad \text{q.e.d.}$$

Again the name deviation is justified. For $a \in G_p(V)$ define

$$\Phi_L[a] = c(L,\kappa) + dd^c \log \|a,\square\|_\kappa^2$$

on $N - E_L[a]$. Obviously $d\Phi_L[a] = 0$ and $\Phi_L[a]$ is a form of class C^∞ and bidegree (1,1) on $N - E_L[a]$. Lemma 4.6 implies $\Phi_L[\infty] = c(L,\ell)$.

Lemma 4.7. Take $a \in G_p(V)$. Then $\Phi_L[a] \geqq 0$ does not depend on the choice of κ and

$$\Phi_L[a] = \varphi^*(\Phi[a]) \qquad \Phi_L[a]^{p+1} = 0.$$

Proof. The map φ is holomorphic on $N - E_L[a]$. Then (4.1), (3.9) and (1.4) imply

$$\Phi_L[a] = c(L,\kappa) + dd^c \log \|\infty,\square\|_\kappa^2 + dd^c \log \|a,\varphi\|^2$$

$$= c(L,\ell) + \varphi^*(dd^c \log \|a,\square\|^2)$$

$$= \varphi^*(\ddot{\omega} + dd^c \log \|a,\square\|^2)$$

$$= \varphi^*(\Phi[a]) \geqq 0.$$

Also $\Phi_L[a]^{p+1} = \varphi^*(\Phi[a]^{p+1}) = \varphi^*(0) = 0$; q.e.d.

Take $a \in G_p(V)$. On N_∞ the Chern-Levine forms are defined by

$$\hat{\Lambda}_L[a]_\kappa = \sum_{\mu=0}^{p} \Phi_L[a]^\mu \wedge c(L,\kappa)^{p-\mu}$$

$$\Lambda_L[a]_\kappa = -\operatorname{Log}\|a,\square\|_\kappa^2\,\hat{\Lambda}_L[a]_\kappa$$

$$\hat{\Lambda}_L[a] = \hat{\Lambda}_L[a]_\ell \geqq 0 \qquad \Lambda_L[a] = \Lambda_L[a]_\ell \geqq 0.$$

They are of bidegree (p,p) and class C^∞. If $c(L,\kappa) \geqq 0$, then $\hat{\Lambda}_L[a]_\kappa \geqq 0$. If also κ is distinguished, then $\Lambda_L[a]_\kappa \geqq 0$.

Lemma 4.8. Take $a \in G_p(V)$. On $N - E_L[a]$, we have

$$dd^c\,\Lambda_L[a]_\kappa = c(L,\kappa)^{p+1} \qquad dd^c\,\Lambda_L[a] = c(L,\ell)^{p+1} = \varphi^*(\ddot{\omega}^{p+1}).$$

Proof.

$$dd^c\,\Lambda_L[a]_\kappa = -\,dd^c\operatorname{Log}\|a,\square\|_\kappa^2\sum_{\mu=0}^{p}\Phi_L(a)^\mu \wedge c(L,\kappa)^{p-\mu}$$

$$= (c(L,\kappa) - \Phi_L[a])\sum_{\mu=0}^{p}\Phi_L[a]^\mu \wedge c(L,\kappa)^{p-\mu}$$

$$= c(L,\kappa)^{p+1} - \Phi_L[a]^{p+1} = c(L,\kappa)^{p+1}$$

q.e.d.

Let I be the integral average over $G_p(V)$. Let $\mathfrak{R}(N)$ be the set of simple points of N. Theorem 1.1 and the previous result imply easily

Theorem 4.9. Take $p \in \mathbb{Z}[0,k)$ and $q \in \mathbb{Z}[0,k]$. Define $s = k - p$. On $\mathfrak{R}(N_\infty)$ we have

$$I(\Phi_L[a]^q) = c(L,\ell)^q$$

$$I(\log\|a,\square\|_\ell^2\,\Phi_L[a]^q) = -\left(\sum_{\mu=1}^{p}\frac{1}{p-q+\mu}\right)c(L,\ell)^q$$

$$I(\hat{\Lambda}_L[a]) = (p+1)\,c(L,\ell)^p$$

$$I(\Lambda_L[a]) = \left(\sum_{\nu=0}^{p}\sum_{\mu=1}^{s}\frac{1}{\nu+\mu}\right)c(L,\ell)^p$$

$$I(\hat{\Lambda}_L[a]_\kappa) = \sum_{\mu=0}^{p} c(L,\ell)^\mu \wedge c(L,\kappa)^{p-\mu}$$

$$I(\Lambda_L[a]_\kappa) = \sum_{\nu=0}^{p}\left(-\log\|\infty,\square\|_\kappa^2 + \sum_{\mu=1}^{s}\frac{1}{\nu+\mu}\right)c(L,\ell)^{p-\nu} \wedge c(L,\kappa)^\nu.$$

If $L = S_0(V^*)^*$ is the hyperplane section bundle over $N = \mathbb{P}(V^*)$ with $\kappa = \ell$, then (3.8) gives an amplification of L. The dual classification map is the identity. Hence

$$\|\infty, x\|_\ell = 1 \qquad \|a,x\|_\ell = \|a,x\| \qquad E_L[a] = \ddot{E}[a]$$

$$\Phi_L[a] = \Phi[a] \qquad c(L,\ell) = \ddot{\omega}$$

$$\hat{\Lambda}_L[a]_\ell = \hat{\Lambda}[a] \qquad \Lambda_L[a]_\ell = \Lambda[a].$$

Let M be a complex space and consider a holomorphic map $f\colon M \to N$. Then f is said to be <u>safe</u> if $f^{-1}(E_L[\infty])$ is thin. If η is ample, f is safe. The holomorphic map f pulls back L to a line bundle L_f over M. Then $M \times V$ is the pullback of $N \times V$ under f. The simplification η pulls back to a simplification $\eta_f\colon M \times V \to L_f$. The following diagrams commute

(4.4)
$$\begin{array}{ccc} L_f & \xrightarrow{\tilde f} & L \\ \Big\downarrow{\scriptstyle\pi} & & \Big\downarrow{\scriptstyle\pi} \\ M & \xrightarrow{f} & N \end{array} \qquad\qquad \begin{array}{ccc} M \times V & \xrightarrow{\eta_f} & L_f \\ \Big\downarrow{\scriptstyle\tilde f} & & \Big\downarrow{\scriptstyle\tilde f} \\ N \times V & \xrightarrow{\eta} & L \end{array}$$

If $L_f = \{(z,u) \in M \times L \mid \varphi(z) = \pi(u)\}$ is the standard model, then $\pi\colon L_f \to M$ and $\tilde f\colon L_f \to L$ are the projections. The pullback $\tilde f = f \times \mathrm{Id}\colon M \times V \to N \times V$ is not the standard model. The map η_f is defined by $\eta_f(z,\mathfrak{v}) = (z, \eta(f(z),\mathfrak{v}))$. Hence $\eta_{f\mathfrak{v}}(z) = (z, \eta_{\mathfrak{v}}(f(z)))$. The extension η in (3.1) pulls back to η_f which is the extension of the pullback $\eta_f\colon M \times V \to L_f$. The hermitian metric κ pulls back to a hermitian metric κ_f along the fibers of L_f and so does ℓ over N_∞ with $\ell = \ell_f$ on the complement of $f^{-1}(E_L[\infty]) = E_{L_f}[\infty]$. Easily the following relations are verified:

(4.5) $$\|a,z\|_{\kappa_f} = \|a,f(z)\|_\kappa$$

(4.6) $$E_{L_f}[a] = f^{-1}(E_L[a])$$

(4.7) $$\|a,z\|_\ell = \|a,f(z)\|_\ell$$

(4.8) $$c(L_f,\kappa_f) = f^*(c(L,\kappa))$$

(4.9) $$c(L_f,\ell) = f^*(c(L,\ell))$$

(4.10) $$\Phi_{L_f}[a] = f^*(\Phi_L[a])$$

(4.11) $$\hat{\Lambda}_{L_f}[a]_{\kappa_f} = f^*(\hat{\Lambda}_L[a]_\kappa)$$

(4.12) $$\Lambda_{L_f}[a]_{\kappa_f} = f^*(\Lambda_L[a]_\kappa)$$

(4.13) $$\hat{\Lambda}_{L_f}[a] = f^*(\hat{\Lambda}_L[a])$$

(4.14) $$\Lambda_{L_f}[a] = f^*(\Lambda_L[a]).$$

The map f is safe, if and only if η_f is semi-ample. If so the dual classification map $\varphi_f\colon M \longrightarrow \mathbb{P}(V)$ is defined, meromorphic and holomorphic on $M_\infty = M - E_{L_f}[\infty] = f^{-1}(N_\infty)$. Then $\varphi_f = \varphi \circ f$ on M_∞. If $a \in G_p(V)$, then

(4.15) $$\varphi_f^{-1}(\ddot{E}[a]) \cap M_\infty = f^{-1}(E_L[a]) \cap M_\infty$$

(4.16) $$\varphi_f^{-1}(E[a]) \subseteq f^{-1}(E_L[a])$$

(4.17) $$\varphi_f^*(\ddot{\omega}) = f^*(c(L,\ell)) \qquad \text{on } M_\infty$$

(4.18) $$\varphi_f^*(\Phi[a]) = f^*(\Phi_L[a])$$

(4.19) $$\varphi_f^*(\hat{\Lambda}[a]) = f^*(\hat{\Lambda}_L[a])$$

(4.20) $$\varphi_f^*(\Lambda[a]) = f^*(\Lambda_L[a])$$

where (4.18) - (4.20) hold on $M_\infty - f^{-1}(E_L[a])$.

The map f is said to be <u>safe of order p</u>, if f is each branch B of M and for each $x \in B \cap f^{-1}(E_L[\infty])$ the estimate

$$\dim B - \dim_x f^{-1}(E_L[\infty]) \cap B > p$$

holds which is the case if and only if η_f is a semi-amplification of order p. If f is <u>safe of order p + 1</u> and if $a \in G_p(V)$, then

(4.21) $$\varphi_f^{-1}(\ddot{E}[a]) = f^{-1}(E_L[a]).$$

If $f\colon M \longrightarrow N$ is a meromorphic map, let F be the closed graph of f over M and let $\vec{f}\colon F \to N$ and $\overleftarrow{f}\colon F \to M$ be the projections. Then f is said to be safe (safe of order p) if and only if $\vec{f}$ is safe (safe of order p).

5. Adaptation

The purpose of adaptation is to exclude degenerate fibers $f^{-1}(E_L[a])$. If L is ample and if f is holomorphic the situation is easy. If L is semi-ample or if f is meromorphic, technical difficulties appear.

a) The incidence space

Let V be a complex vector space of dimension $k + 1 \geqq 1$. Take $p \in \mathbb{Z}[0,k)$. Define $d(p,k) = (k - p)(p + 1)$. Then the incidence space

$$\mathbb{F} = \{(x,a) \in \mathbb{P}(V^*) \times G_p(V) \mid x \in \ddot{E}[a]\}$$

is a compact, connected, smooth, complex submanifold of dimension $k - p + 1 + d(p,k)$ in $\mathbb{P}(V^*) \times G_p(V)$. The projections $\pi\colon \mathbb{F} \to \mathbb{P}(V^*)$ and $\sigma\colon \mathbb{F} \to G_p(V)$ are surjective, regular, holomorphic maps of fiber dimensions $(k - p - 1)(p + 1)$ and $(k - p - 1)$ respectively. The fibers are connected with $\sigma^{-1}(a) = \ddot{E}(a) \times \{a\}$. (For instance, see [76] Proposition 2.1.)

Lemma 5.1. Let V be a complex vector space of dimension $k + 1 \geqq 1$. Take $p \in \mathbb{Z}[0,k]$. Let M be a complex space of pure dimension m with $q = m - p - 1 \geqq 0$. Let $f\colon M \longrightarrow \mathbb{P}(V^*)$ be a meromorphic map. Then the incidence space

$$\mathbb{F}(f) = \{(x,a) \in M \times G_p(V) \mid x \in f^{-1}(\ddot{E}[a])\}$$

is a pure $(q + d(p,k))$-dimensional analytic subset of $M \times G_p(V)$. Let $\tilde{f}\colon \mathbb{F}(f) \to G_p(V)$ and $\tilde{\pi}\colon \mathbb{F}(f) \to M$ be the projections. Then $\tilde{\pi}$ is proper. If $a \in G_p(V)$, then

$$\hat{f}^{-1}(a) = f^{-1}(\ddot{E}[a]) \times \{a\}.$$

Let I_f be the indeterminacy of f. Define $M_0 = M - I_f$. Then $\mathbb{F}(f) \cap (M_0 \times G_p(V))$ is open and dense in $\mathbb{F}(f)$.

Proof. Let $\hat{M}$ be the graph of f. Let $\vec{f}\colon \hat{M} \to \mathbb{P}(V^*)$ and $\overleftarrow{f}$ be the projections. Take $\mathbb{F}$, π, σ as before. Let $\hat{\pi}\colon \mathbb{F}(\vec{f}) \to \hat{M}$ and $\psi\colon \mathbb{F}(\vec{f}) \to G_p(V)$ be the projections. Define $\varphi\colon \mathbb{F}(\vec{f}) \to \mathbb{F}$ by $\varphi(x,y,a) = (\vec{f}(x,y),a) = (y,a)$. Define $\tilde{\gamma}\colon \mathbb{F}(\vec{f}) \to M \times G_p(V)$ by $\tilde{\gamma}(x,y,a) = (x,a)$. Here $(x,y) \in \hat{M}$ and $y \in \ddot{E}[a]$. Hence $x \in f^{-1}(\ddot{E}[a])$. Therefore $\tilde{\gamma}(x,y,a) \in \mathbb{F}(f)$. Let $\gamma\colon \mathbb{F}(\vec{f}) \to \mathbb{F}(f)$ be the restriction of $\tilde{\gamma}$. If $(x,a) \in \mathbb{F}(f)$, then $x \in f^{-1}(\ddot{E}[a])$ and $y \in \ddot{E}[a]$ exists with $(x,y) \in \hat{M}$. Hence $(x,y,a) \in \mathbb{F}(\vec{f})$ with $\gamma(x,y,a) = (x,a)$. Therefore γ is surjective. The commutative diagram 5.1 is constructed.

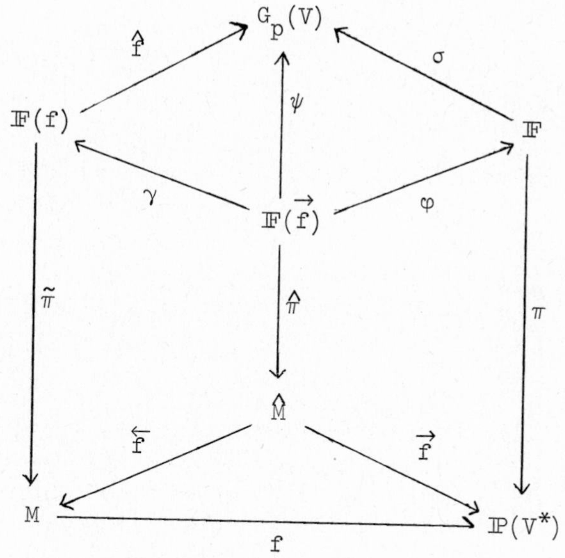

Diagram 5.1.

Clearly, $(\mathbb{F}(\vec{f}),\varphi,\hat{\pi})$ is the relative product of $(\vec{f},\pi)$. Hence $\mathbb{F}(\vec{f})$ is an analytic subset of pure dimension $m + (k - p - 1)(p + 1) = q + d(p,k)$ of $\hat{M} \times G_p(V)$ and $\hat{\pi}$ is proper. Since $\tilde{\gamma}$ is proper, $\mathbb{F}(f)$ is analytic.

Let I_f be the indeterminacy of f. Then $\tilde{I}_f = \overleftarrow{f}^{-1}(I_f)$ is thin analytic in $\hat{M}$. Define $M_0 = M - I_f$ and $\hat{M}_0 = \hat{M} - \tilde{I}_f$. Then $f\colon \hat{M}_0 \to M_0$ is biholomorphic. Because π is locally trivial, so is $\hat{\pi}$. Hence $\hat{I}_f = \hat{\pi}^{-1}(\tilde{I}_f)$ is thin analytic in $\mathbb{F}(\vec{f})$. Then $I_f^0 = \gamma(\hat{I}_f)$ is analytic in $\mathbb{F}(f)$. Define $\tilde{L}_0 = \mathbb{F}(\vec{f}) - \hat{I}_f$ and $L_0 = \mathbb{F}(f) - I_f^0$. Then

$$\tilde{L}_0 = \{(x,f(x),a) \in M_0 \times \mathbb{P}(V^*) \times G_p(V) \mid x \in f^{-1}(\ddot{E}[a])\}$$

$$L_0 = \{(x,a) \in M_0 \times G_p(V) \mid x \in f^{-1}(\ddot{E}[a])\}.$$

Hence $\gamma\colon \tilde{L}_0 \to L_0$ is biholomorphic and L_0 has pure dimension $q + d(p,k)$. Since γ is proper and $\tilde{L}_0$ is dense, $\overline{L}_0 = \gamma(\mathbb{F}(\vec{f})) = \mathbb{F}(f)$. Hence $L_0 = (M_0 \times G_p(V)) \cap \mathbb{F}(f)$ is dense in $\mathbb{F}(f)$. Therefore $\mathbb{F}(f)$ has pure dimension $q + d(p,k)$. Since $\overleftarrow{f}$ and $\hat{\pi}$ are proper, also $\tilde{\pi}$ is proper; q.e.d.

b) Adaptation for holomorphic maps

Throughout section b), the following assumptions shall be made: Let L be a holomorphic line bundle over the complex space N. A semi-amplification $\eta\colon N \times V \to L$ with $\dim V = k + 1$ is given. Let M be a complex space of pure dimension $m > 0$. Consider a holomorphic map $f\colon M \to N$. For $a \in G_p(V)$, denote $f_a = f^{-1}(E_L[a])$. Take $p \in \mathbb{Z}[0,k]$ with $q = m - p - 1 \geqq 0$.

Lemma 5.2. If $a \in G_p(V)$ and $x \in f_a$, then $\dim_x f_a \geqq q$.

Proof. Let v be a holomorphic frame of L over an open neighborhood U of f(x). Define $\tilde{U} = f^{-1}(U)$. Determine $\mathfrak{w}$ by (3.4). Take a base $\mathfrak{a}_0,\dots,\mathfrak{a}_k$ of V such that $a = \mathbb{P}(\mathfrak{a}_0 \wedge \cdots \wedge \mathfrak{a}_p)$. Let $\alpha_0,\dots,\alpha_k$ be the dual base. Then $\mathfrak{w} = w_0\alpha_0 + \dots + w_k\alpha_k$. Define $\mathfrak{v} = (w_0 \circ f,\dots,w_p \circ f)$. Then $\tilde{U} \cap f_a = \mathfrak{v}^{-1}(0)$. Hence $\dim_x f_a \geqq q$; q.e.d.

The incidence space of f for L and p is defined by

$$(5.2) \qquad \mathbb{F}_L(f) = \{(x,a) \in M \times G_p(V) \mid x \in f_a\}$$

Let $\pi: \mathbb{F}_L(f) \longrightarrow M$ and $\hat{f}: \mathbb{F}_L(f) \longrightarrow G_p(V)$ be the projections. Then

$$(5.3) \qquad \hat{f}^{-1}(a) = f_a \times \{a\}$$

<u>Lemma 5.3.</u> $\mathbb{F}_L(f)$ is analytic in $M \times G_p(V)$. Also

$$(5.4) \qquad \dim_z \mathbb{F}_L(f) \geqq q + d(p,k)$$

for $z \in \mathbb{F}_L(f)$. Let φ be the dual classification map. If f is safe of order p + 1, the map $\varphi_f = \varphi \circ f: M \longrightarrow \mathbb{P}(V^*)$ is defined and meromorphic and $\mathbb{F}_L(f) = \mathbb{F}(\varphi_f)$ has pure dimension $q + d(p,k)$. Observe

$$(5.5) \qquad \mathbb{F}(\varphi_f) = \{(x,a) \in M \times G_p(V) \mid x \in (\varphi \circ f)^{-1}(\ddot{E}[a])\}.$$

<u>Proof.</u> Take $(x,a) \in M \times G_p(V)$. Let v be a holomorphic frame of L over an open neighborhood U of f(x). Define $\tilde{U} = f^{-1}(U)$. Determine $\mathfrak{w}$ by (3.4) for $\varphi \circ f$. An open neighborhood Y of a in $G_p(V)$ and holomorphic vector functions $\mathfrak{g}_\mu: Y \to V$ exist such that $\mathfrak{g}(y) = \mathfrak{g}_0(y) \wedge \cdots \wedge \mathfrak{g}_p(y) \neq 0$ and $y = \mathbb{P}(\mathfrak{g}(y))$ for all $y \in V$. Define $g_\mu: \tilde{U} \times Y \to \mathbb{C}$ by $g_\mu(x,y) = (\mathfrak{w}(x), \mathfrak{g}_\mu(y))$. Then (3.2) and (3.4) imply

$$\mathbb{F}_L(f) \cap (\tilde{U} \times Y) = \bigcap_{\mu=0}^{p} g_\mu^{-1}(0).$$

Therefore $\mathbb{F}_L(f)$ is analytic with

$$\dim_{(x,a)} \mathbb{F}_L(f) \geqq m + d(p,k)) - p - 1 = q + d(p,k).$$

The rest follows from (4.21) and Lemma 5.1; q.e.d.

The map f is said to be <u>truly adapted</u> to $a \in G_p(V)$ for L at $x \in M$ if $\dim_x f_a = q$. Then $x \in f_a$. The map f is said to be <u>adapted</u> to a ffor L at x if either $x \in M - f_a$ of if f is truly adapted to a at x. The map f is said to be <u>adapted</u> to a for L, if f is adapted to a for L at all $x \in M$.

<u>Lemma 5.4.</u> The map f is truly adapted to $a \in G_p(V)$ for L at x if and only if $z = (x,a) \in \mathbb{F}_L(f)$ with $\dim_z \mathbb{F}_L(f) = q + (k - p)(p + 1)$ and if $\hat{f}$ is open in a neighborhood of z.

<u>Proof.</u> a) <u>Assume that f is truly adapted to a for L at x.</u> Then $z \in \mathbb{F}_L(f)$. The properties of the pseudo-rank (see [2]) imply

$$\begin{aligned} d(p,k) &\geqq \widetilde{\operatorname{rank}}_z \hat{f} = \dim_z \mathbb{F}_L(f) - \dim_z \hat{f}^{-1}(a) \\ &\geqq q + d(p,k) - \dim_x f_a = d(p,k). \end{aligned}$$

Hence $\dim_z \mathbb{F}_L(f) = q + d(p,k)$. By Lemma 5.3 an open neighborhood U of z exists such that U has pure dimension $q + d(p,k)$. Hence $\operatorname{rank}_z \hat{f} = \widetilde{\operatorname{rank}}_z \hat{f} = d(p,k)$ and $\hat{f}$ is open at z by Remmert.

b) <u>Assume that $\hat{f}$ is open at $z = (x,a) \in \mathbb{F}_L(f)$ with $\dim_z \mathbb{F}_L(f) = q + d(p,k)$.</u> Again an open, pure $(q + d(p,k))$ - dimensional neighborhood U of z exists, such that $\hat{f}$ is open on U. Hence $\hat{f}$ is q - fibering on U. Therefore $\dim_x f_a = q$ by (5.3); q.e.d.

<u>Remark to Lemma 5.4.</u> If f is truly adapted to a for L at x, then $\mathbb{F}_L(f)$ has pure dimension $q + d(p,k)$ in some neighborhood of (x,a).

If $K \subseteq M$, define $\mathfrak{A}_p(K,L,f)$ as the set of all $a \in G_p(V)$ such that f is adapted to a for L at each $x \in K$. Let $\mathfrak{A}_p^t(K,L,f)$ be the set of all $a \in G_p(V)$ such that $K \cap f_a \neq \emptyset$ and such that f is (truly) adapted to a for L at each $x \in K \cap f_a$. Also, let $\mathring{\mathfrak{A}}_p^t(K,L,f)$ be the set of all $a \in \mathfrak{A}_p^t(K,L,f)$ with $f_a \cap \operatorname{Int} K_a \neq \emptyset$. Obviously

$$\mathring{\mathfrak{A}}_p^t(K,L,f) \subseteq \mathfrak{A}_p^t(K,L,f) \subseteq \mathfrak{A}_p(K,L,f).$$

Proposition 5.5. If K is a compact subset of M, then $\mathfrak{A}_p(K,L,f)$ and $\mathring{\mathfrak{A}}^t_p(K,L,f)$ are open in $G_p(V)$.

Proof. Define analytic subsets of $\mathbb{F}_L(f)$ by

$$T = \{z \in \mathbb{F}_L(f) \mid \operatorname{rank}_z \hat{f} < d(p,k)\}$$

$$D = \{z \in \mathbb{F}_L(f) \mid \dim_z \mathbb{F}_L(f) > q + d(p,k)\}.$$

The projection $\pi: \mathbb{F}_L(f) \longrightarrow M$ is proper. Hence $\tilde{K} = \pi^{-1}(K)$ is compact. Then $(T \cup D) \cap \tilde{K} = \tilde{K}_1$ and $K_2 = \hat{f}(\tilde{K}_1)$ are compact. Lemma 5.4 implies

$$K_2 = G_p(V) - \mathfrak{A}_p(K,L,f).$$

Thus $\mathfrak{A}_p(K,L,f)$ is open.

Take $a \in \mathring{\mathfrak{A}}^t_p(K,L,f)$. The restriction $\pi_a = \pi: \hat{f}^{-1}(a) \longrightarrow f_a$ is biholomorphic. Hence $\tilde{K} \cap \hat{f}^{-1}(a) = \pi_a^{-1}(f_a \cap K) \neq \emptyset$. If $x \in K \cap f_a$, then f is adapted to a for L at x. Hence $(x,a) \notin T \cup D$. Therefore $\tilde{K}_1 \cap \hat{f}^{-1}(a) = \emptyset$. An open neighborhood U of $\tilde{K} \cap \hat{f}^{-1}(a)$ exists with $U \cap (T \cup D) = \emptyset$ such that $\hat{f}: U \longrightarrow G_p(V)$ is open. An open neighborhood W of a in $G_p(V)$ exists such that $\hat{f}^{-1}(W) \cap \tilde{K} \subseteq U$. Now, $K_0 = \pi^{-1}(\text{Int } K) \cap U$ is open. Since $a \in \mathring{\mathfrak{A}}^t_p(K,L,f)$, there exists $x_0 \in (\text{Int } K) \cap f_a$ such that L is adapted to a at x_0. Then $(x_0,a) \in K_0$. Hence $K_0 \neq \emptyset$. Also $a = \hat{f}(x_0,a) \in \hat{f}(K_0)$. Then $W_1 = W \cap \hat{f}(K_0)$ is an open neighborhood of a. Take $b \in W_1$. Then $(x,b) \in K_0 \cap \hat{f}^{-1}(b)$ exists which implies $x \in (\text{Int } K) \cap f_b$. Hence $(\text{Int } K) \cap f_b \neq \emptyset$. Take any $z \in K \cap f_b$; then $(z,b) \in \tilde{K} \cap \hat{f}^{-1}(b) \subseteq U$ since $b \in W$. Therefore $\hat{f}$ is open at (z,b) and $(z,b) \notin T \cup D$. Consequently, f is adapted to b for L at z. Hence $b \in \mathring{\mathfrak{A}}^t_p(K,L,f)$. So $W_1 \subseteq \mathring{\mathfrak{A}}^t_p(K,L,f)$. The set $\mathring{\mathfrak{A}}^t_p(K,L,f)$ is open; q.e.d.

Proposition 5.6. Assume f is safe of order p + 1. Then the projection $\pi: \mathbb{F}_L(f) \longrightarrow M$ is surjective. The complement of $\mathfrak{A}_p(M,L,f)$ is almost thin in $G_p(V)$.

Proof. By Lemma 5.3, $\mathbb{F}_L(f)$ has pure dimension $q + d(p,k)$. Define T and D as in the proof of Proposition 5.5. Then $D = \emptyset$ and $\hat{f}(T) = G_p(V) - \mathfrak{A}_p(M,L,f)$ is almost thin by [2] Lemma 1.30. Let φ be the dual classification map of L. Then $\varphi \circ f: M \longrightarrow \mathbb{P}(V^*)$ is meromorphic with I as its indeterminacy. Then $I \subseteq f_\infty$ where $\infty \in G_k(V)$. Define $M_0 = M - I$ and $\tilde{I} = \pi^{-1}(I)$. By Lemma 5.1, $\tilde{I}$ is thin in $\mathbb{F}_L(f) = \mathbb{F}(\varphi \circ f)$. By diagram 5.1, $\pi: \mathbb{F}_L(f) - \tilde{I} \longrightarrow M_0$ is surjective with connected manifolds of dimension $(k - p - 1)(p + 1)$ as fibers. Since π is proper, $\pi(\mathbb{F}_L(f))$ is analytic. Because $\pi(\mathbb{F}_L(f))$ contains M_0, the map π is surjective; q.e.d.

The holomorphic map f is said to be almost adapted for L of order p if for each branch B of M there exist $a \in G_p(V)$ and $x \in B \cap f_a - f_\infty$ such that f is truly adapted to a for L at x.

Theorem 5.7. Assume f is safe of order p + 1. Then f is almost adapted for L of order p if and only if $\hat{f}$ has strict rank $d(p,k)$. Let K be compact in M with non-empty interior. If f is almost adopted to L of order p, then $\mathring{\mathfrak{A}}^t_p(K,L,f) \neq \emptyset$.

Proof. By Lemma 5.3. $\mathbb{F}_L(f)$ has pure dimension $q + d(p,k)$. Define T, $D = \emptyset$, φ, I, M_0 and $\tilde{I}$ as in the proof of Proposition 5.6. Then T is thin if and only if $\hat{f}$ has strict rank $d(p,k)$.

Assume $\hat{f}$ has strict rank $d(p,k)$. Let B be a branch of M. An open subset $U \neq \emptyset$ of $B - f_\infty$ exists. Then $\tilde{U} = \pi^{-1}(U) \neq \emptyset$ is open. Since T is thin, $(x,a) \in \tilde{U} - T$ exists. Then $x \in B \cap f_a - f_\infty$ and f is truly adapted to a for L at x. The map f is almost adapted for L of order p.

Assume f is almost adapted for L of order p. Let C be a branch of $\mathbb{F}_L(f)$. Then $C_0 = C - \tilde{I}$ is a branch of $\mathbb{F}_L(f) - \tilde{I}$, since $\tilde{I}$ is thin. Because $\pi_0: C_0 \longrightarrow M_0$ has pure fiber dimension $(k - p - 1)(p + 1)$ and is proper, $B_0 = \pi(C_0)$ is an irreducible analytic subset of dimension m in M_0. Hence B_0 is a branch of M_0 and $B = \overline{B}_0$ is a branch of M. The fibers of π over B_0 are connected, complex manifolds of dimension $(p + 1)(k - p - 1)$. Hence $C_0 = \pi^{-1}(B_0)$. Now, $a \in G_p(V)$ and $x \in B \cap f_a - f_\infty \subseteq B_0$ exist such that f is truly adapted to a for L at x. Hence $(x,a) \in C_0 - T$. Therefore $C - T \neq \emptyset$. The analytic set T is thin in $\mathbb{F}_L(f)$. Hence $\hat{f}$ has strict rank $d(p,k)$.

Let K be compact in M with Int $K \neq 0$. Again assume that f is almost adapted for L of order p. Since π is surjective and $T \cup \tilde{I}$ is thin, a point $(x_0,a_0) \in \pi^{-1}(\text{Int } K) - (T \cup \tilde{I}) = W$ exists. Then $\hat{f}$ is open in an open neighborhood U of (x_0,a_0) with $U \subseteq W$. The image $\hat{f}(U)$ is open. By Proposition 5.6 $a \in \hat{f}(U) \cap \mathfrak{A}_p(M,L,f)$ exists. Take $(z,a) \in U$. Then $z \in \text{Int } K \cap f_a \neq \emptyset$. Also $a \in \mathfrak{A}_p(M,L,f)$ implies that f is adapted to a for L at all $x \in M$, especially at all $x \in K \cap f_a$. Therefore $a \in \mathring{\mathfrak{A}}^t_p(K,L,f) \neq \emptyset$;q.e.d.

Let φ: $N \rightarrow \mathbb{P}(V^*)$ be the dual classification map of L. Assume f: $M \longrightarrow N$ is safe of order p + 1 for L. Then $\varphi_f = \varphi \circ f: M \rightarrow \mathbb{P}(V^*)$ is meromorphic and holomorphic on $M_\infty = M - f_\infty$. Let $\hat{M}$ be the closed graph of φ_f and let $\overleftarrow{\varphi}_f: \hat{M} \longrightarrow M$ and $\overrightarrow{\varphi}_f: \hat{M} \longrightarrow \mathbb{P}(V^*)$ be the projections. Take $a \in G_p(V)$. Then f is <u>said to be strictly adapted to a for L</u>, if $\overrightarrow{\varphi}_f$ is adapted to a for $S_0(V^*)^*$. Since $S_0(V^*)^*$ is ample, $\overrightarrow{\varphi}_f$ is safe of order p + 1. Hence the set $\mathfrak{A}_p(M,L,f) = \mathfrak{A}_p(M,S_0(V^*)^*,\overrightarrow{\varphi}_f)$ of all $a \in G_p(V)$ such that f is strictly adapted to a for L has a complement of measure zero.

<u>Lemma 5.8.</u> If f is strictly adapted to $a \in G_p(V)$ for L, then f is adapted to a for L.

<u>Proof.</u> By assumption $S = \overrightarrow{\varphi}_f^{-1}(\ddot{E}[a])$ is empty or pure q-dimensional. Now, (4.21) implies $\overleftarrow{\varphi}_f(S) = f^{-1}(E_L[a]) = f_a$. Hence dim $f_a \leqq q$. Lemma 5.1 implies that $f_a = \emptyset$ or that f_a has pure dimension q. Hence f is adapted to a for L,q.e.d.

Lemma 5.9. If $p = 0$, then f is strictly adapted to a for L, if and only is f is adapted to a for L.

Proof. Assume f is adapted to a for L. Then f_a is thin with $\overleftarrow{\varphi}_f^{-1}(f_a) \equiv \overrightarrow{\varphi}_f^{-1}(\ddot{E}[a]) = S$ by (4.16). Hence $\dim S \leqq m - 1$. By Lemma 5.1, S is either empty or pure $(m - 1)$ dimensional. Hence f is strictly adapted to a for L;q.e.d.

Assume $M = N$ and let f be the identity. Then L is said to be (truly) adapted to a (at x), if the identity is (truly) adapted to a L(at x). If the identity is almost adapted to $\acute{L}$ of order p, call L almost adapted of order p. Also denote

$$\mathfrak{A}_p(K,L) = \mathfrak{A}_p(K,L,Id)$$

$$\mathfrak{A}_p^t(K,L) = \mathfrak{A}_p^t(K,L,Id)$$

$$\mathring{\mathfrak{A}}_p^t(K,L) = \mathring{\mathfrak{A}}_p^t(K,L,Id).$$

If the identity is strictly adapted to a for L, call L strictly adapted to a.

c) Adaption for meromorphic maps

Let $f: M \to N$ be a safe meromorphic map. Let F be the graph of f and let $\overleftarrow{f}: F \longrightarrow M$ and $\overrightarrow{f}: F \longrightarrow N$ be the projections. Then f is said to be truly adapted to $a \in G_p(V)$ for L at $x \in M$, if $x \in f_a = f^{-1}(E_L[a])$ and if $\overrightarrow{f}$ is adapted to a for L at every $y \in \overleftarrow{f}^{-1}(x)$. The map f is said to be adapted to a for L at x if either $x \notin f^{-1}(E_L[a])$ or if $x \in f^{-1}(E_L[a])$ and f is truly adapted to a for L at x. The map f is said to be adapted to a for L, if f is adapted to a for L at all $x \in M$ Also f is said to be strictly adapted to a for L if $\overrightarrow{f}$ is strictly adapted to a for L. The map f is said to be almost adapted for L, if $\overrightarrow{f}$ is almost adapted for L. If $\emptyset \neq K \subseteq N$, then $\mathfrak{A}_p(K,L,f)$, $\mathfrak{A}_p^t(K,L,f)$ and $\mathring{\mathfrak{A}}_p^t(K,L,f)$ are defined as before. If $\hat{K} = \overleftarrow{f}^{-1}(K)$, then

$$\mathfrak{A}_p(K,L,f) = \mathfrak{A}_p(\hat{K},L,\vec{f})$$
$$\mathfrak{A}_p^t(K,L,f) = \mathfrak{A}_p^t(\hat{K},L,\vec{f})$$
$$\mathring{\mathfrak{A}}_p^t(K,L,f) = \mathring{\mathfrak{A}}_p^t(\hat{K},L,\hat{f}).$$

If K is compact, then $\hat{K}$ is compact. Hence Proposition 5.5, Proposition 5.6, Theorem 5.7, Lemma 5.8 and Lemma 5.9 hold for safe meromorphic maps.

6. Multiplicities

a) Multiplicities of holomorphic maps

Let M and N be complex spaces of pure dimensions m and n respectively with $q = m - n \geqq 0$. Assume that N locally irreducible. Let $f: M \longrightarrow N$ be a q-fibering holomorphic map. For $a \in M$, abbreviate $F_a = f^{-1}(f(a))$. According to Tung [87] the multiplicity $\nu_f(a)$ of f at a is defined as follows:

At first consider $q = 0$. An open neighborhood U of a is called distinguished if $\overline{U}$ is compact and if $F_a \cap \overline{U} = \{a\}$. Then

$$1 \leqq \nu_f(a) = \limsup_{z \to a} \# F_z \cap U$$

is an integer independent of U which is called the multiplicity of f at a. Now consider $q > 0$. A holomorphic map $\varphi: U_\varphi \longrightarrow \mathbb{C}^q$ on an open neighborhood U_φ of a is called a slicing map of f at a if $(f,\varphi): U_\varphi \longrightarrow N \times \mathbb{C}^q$ is light. Then $\nu_{(f,\varphi)}(a)$ is defined. Let $\Phi_a^0(f)$ be the set of all slicing maps of f at a. Let e_a be the embedding dimension of M at a. Then $\varphi \in \Phi_a^0(f)$ is said to be regular if there exists a patch

$$\alpha\colon U_\varphi = U_\alpha \longrightarrow U'_\alpha \subseteq G_\alpha \subseteq \mathbb{C}^{e_a}$$

and a regular holomorphic map $\tilde{\varphi}\colon G_\alpha \longrightarrow \mathbb{C}^q$ such that $\tilde{\varphi}\circ\alpha = \varphi \mid U_\alpha$. Observe U'_α is analytic in the open set G_α and α is biholomorphic. The map $\tilde{\varphi}$ is regular, if its Jacobian has constant rank q. The set $\Phi^1_a(f)$ of all regular slicing maps of f at a is not empty. For $\varphi \in \Phi^0_a(f)$ the restriction $\psi = \varphi \mid F_a \cap U_\varphi$ is light at a. Hence $\nu^f_\varphi(a) = \nu_\psi(a)$ is defined. Determine

$$d_a = d_a(f) = \operatorname{Min}\, \{\nu^f_\varphi(a) \mid \varphi \in \Phi^1_a(f)\}.$$

Then $\varphi \in \Phi^1_a(f)$ is called <u>minimal</u> if $\nu^f_\varphi(a) = d_a(f)$. Let $\phi^2_a(f)$ be the set of all minimal slicing maps. The <u>multiplicity of f at a</u> is defined by

$$\nu_f(a) = \operatorname{Min}\, \{\nu_{(f,\varphi)}(a) \mid \varphi \in \Phi^2_a(f)\}.$$

If $b \in N$, the <u>b-multiplicity of f at a</u> is defined by $\nu^b_f(a) = 0$ if $f(a) \neq b$ and by $\nu^b_f(a) = \nu_f(a)$ if $b = f(a)$.

Although the definition is complicated, the multiplicity has many fine properties, see Tung [87]. For instance, ν_f is locally constant on the set $\mathfrak{R}(F_a)$ of simple points of F_a for each $a \in N$. The fiber integral using the multiplicity ν_f is continuous. The following Rouché theorem holds.

<u>Rouché's Theorem.</u> Assume $q = 0$. Let Q be a locally connected, connected Hausdorff space. Let $f\colon M \times Q \longrightarrow N$ be continuous. For each $t \in Q$, the map $f_t\colon M \longrightarrow N$ defined by $f_t(x) = f(x,t)$ is holomorphic. Let $G \neq \phi$ be open and relative compact in M. Assume $N_0 \neq \phi$ is open and connected in N with $f_t^{-1}(N_0) \cap \partial G = \phi$ for all $t \in Q$. Then a

non-negative integer s exists such that $s = \sum_{x \in G} \nu^y_{f_t}(x)$ for all $(t,y) \in Q \times N_0$.

The following result is not contained in Tung [87] but is essential to the definition of the multiplicity of sections in holomorphic vector bundles.

Theorem 6.1. Let M be a complex space of pure dimension $m > 0$. Let $f: M \longrightarrow \mathbb{C}^n$ and $h: M \longrightarrow \mathbb{C}^n$ be holomorphic maps with $q = m - n \geqq 0$. Assume that f is q-fibering. Let $g: M \longrightarrow \mathbb{C}^{n^2}$ be a holomorphic matrix function such that $h = f \cdot g$. Take $a \in M$ with $f(a) = 0$. Assume $\det g(a) \neq 0$. Then h is q-fibering in a neighborhood of a and $\nu_h(a) = \nu_f(a)$.

Proof. Trivally h is q-fibering at a. W.l.o.g, h can be assumed to be q-fibering on M and $\det g(x) \neq 0$ for all $x \in M$. Moreover $g(a) = I$ can be assumed to be the unit matrix.

1.Case: $q = 0$, that is $m = n$. Define the holomorphic matrix function G on $M \times \mathbb{C}$ by $G(x,t) = (1 - t)\, I + tg(x)$. Then $G(a,t) = I$ for $t \in \mathbb{C}$. Open, connected neighborhoods M_0 of a and U of $\mathbb{R}[0,1]$ exist with $\det G(x,t) \neq 0$ for all $(x,t) \in M_0 \times U$ and such that $f^{-1}(0) \cap M_0 = \{a\}$. Define $F: M_0 \times U \longrightarrow \mathbb{C}^m$ by $F(x,t) = f(x)G(x,t)$ and $F_t: M_0 \longrightarrow \mathbb{C}^m$ by $F_t(x) = F(x,t)$. Then $F_0 = f \mid M_0$ and $F_1 = h \mid M_0$. Define $\tilde{F}$ by $\tilde{F}(x,t) = (F(x,t),t)$ for all $(x,t) \in M_0 \times U$.
Then $\tilde{F}^{-1}(0,t) = \{(a,t)\}$ for all $t \in U$. Hence open neighborhoods M_1 of a in M_0 and U_1 of $\mathbb{R}[0,1]$ in U exist such that $\tilde{F}$ is light on $M_1 \times U_1$. Then $F_t: M_1 \longrightarrow \mathbb{C}^m$ is light and $F_t^{-1}(0) = \{a\}$ for each $t \in U_1$. Take an open connected neighborhood M_2 of a such that $\overline{M}_2$ is compact and contained in M_1. Then $F_f^{-1}(0) \cap \partial M_2 = \emptyset$ for all $t \in U_1$. Let U_2 be an open, connected neighborhood of $\mathbb{R}[0,1]$ such that $\overline{U}_2$ is compact and contained in U_1. An open connected neighborhood N of $0 \in \mathbb{C}^m$ exists such that $F_t^{-1}(N) \cap \partial M_2 = \emptyset$ for all $t \in \overline{U}_2$.

Rouché's theorem implies that

$$\sum_{x \in M_2} \nu^y_{F_t}(x) = s(t,y)$$

is a constant function of (t,y) in $U_2 \times N$. Therefore $\nu_f(a) = s(0,0) = s(1,0) = \nu_f(b)$. The first case is proved.

2.Case: $q > 0$ that is $m > n$. Obviously, $F = f^{-1}(0) = h^{-1}(0)$. Take $\varphi \in \Phi^1_a(f)$. Define $(f,\varphi) = \psi$ and $(h,\varphi) = \chi$. Then ψ is light. Hence $\psi^{-1}(\psi(a)) = \chi^{-1}(\chi(a))$ is zero dimensional. An open neighborhood U^*_φ of a in U_φ exists such that $\chi \mid U_\varphi$ is light which implies $\varphi \mid U^*_\varphi \in \Phi^1_a(h)$.

Take $\varphi \in \Phi^2_a(f)$. An open neighborhood U^*_φ of a in U_φ exists such that $\varphi \mid U^*_\varphi \in \Phi^1_a(h)$. Define $\rho = f \mid F \cap U_\varphi$. The 1. case implies

$$d_a(h) \leqq \nu^h_\varphi(a) = \nu_\rho(a) = \nu^f_\varphi(a) = d_a(f).$$

By symmetry, $d_a(f) \leqq d_a(h)$. Together $d_a(f) = d_a(h)$ is obtained.

Take $\varphi \in \Phi^2_a(f)$ with $\nu_f(a) = \nu_{(f,\varphi)}(a)$. An open neighborhood U^*_ρ of a in U_φ exists such that $\varphi | U^* \in \Phi^1_a(h)$. Define $\rho = f \mid F \cap U_\varphi$. Then $\nu^h_g(a) = \nu_\rho(a) = \nu^f_\varphi(a) = d_a(f) = d_a(h)$. Hence $\varphi \mid U^*_\varphi \in \Phi^2_a(h)$. Let I be the unit matrix of q lines. Then

$$(h,\varphi) = (f,\varphi)\begin{pmatrix} g & 0 \\ 0 & I \end{pmatrix}$$

$$\det \begin{pmatrix} g(a) & 0 \\ 0 & I \end{pmatrix} = \det g(a) \neq 0.$$

Case 1 implies $\nu_f(a) = \nu_{(f,\varphi)}(a) = \nu_{(h,\varphi)}(a) \geqq \nu_h(a)$. By symmetry $\nu_f(a) \leqq \nu_h(a)$. Consequently $\nu_f(a) = \nu_h(a)$;q.e.d.

Consider a meromorphic map $f: M \longrightarrow N$. Let $\hat{M}$ be the closed graph of f and let $\vec{f}: \hat{M} \longrightarrow N$ and $\overleftarrow{f}: \hat{M} \longrightarrow M$ be the projections. If $b \in N$, then $\vec{f}^{-1}(b) = f^{-1}(b) \times \{b\}$. The restriction

$$\overleftarrow{f}_b = \overleftarrow{f}: \vec{f}^{-1}(b) \longrightarrow f^{-1}(b)$$

is biholomorphic. Hence f is q-fibering if and only if $\vec{f}$ is q-fibering. If $a \in f^{-1}(b)$, define $\nu_f^b(a) = 0$. If $a \in f^{-1}(b)$ and if $\dim_a f^{-1}(b) = q$, then $\vec{f}$ is q-fibering in a neighborhood of (a,b) and $\nu_f^b(a) = \nu_{\vec{f}}(a,b)$ is defined. If $f^{-1}(b)$ has pure dimension q, then $\nu_f^b \circ \overleftarrow{f}_b = \nu_{\vec{f}}^b$ on $\vec{f}^{-1}(b)$ and ν_f^b is locally constant on the set of simple points of $f^{-1}(b)$. Notice ν_f is not defined at points of indetermination.

Let $\mathbb{P}_1 = \mathbb{P}(\mathbb{C}^2)$ be the Riemann sphere. Identify $\mathbb{P}(z,1) = z$ and $\mathbb{P}(1,0) = \infty$. Then a meromorphic function f on M is a meromorphic map $f: M \to \mathbb{P}_1$ such that $f^{-1}(\infty)$ is thin. If $b \in \mathbb{P}_1$ and if $f^{-1}(b)$ is thin, then $f^{-1}(b)$ is empty or pure (m-1) - dimensional. Hence ν_f^b is defined. The pole multiplicity ν_f^∞ is always defined. If $f^{-1}(0)$ is thin, $\theta_f = \nu_f^0 - \nu_f^\infty$ is called the <u>divisor multiplicity of f.</u>

Now, Theorem 6.1 can be extended to meromorphic functions.

<u>Lemma 6.2.</u> Let f be a meromorphic function on the pure m-dimensional complex space M such that $f^{-1}(0)$ is thin. Let h be a holomorphic function on M with $h(x) \neq 0$ for all $x \in M$. Define $g = hf$. Then $\nu_g^0 = \nu_f^0$ and $\nu_g^\infty = \nu_f^\infty$ and $\theta_g = \theta_f$.

<u>Proof</u>. Let F and G be the closed graphs of f and g respectively. Let $\overleftarrow{f}: F \longrightarrow M$ and $\vec{f}: F \longrightarrow \mathbb{P}_1$ and $\overleftarrow{g}: G \longrightarrow M$ and $\vec{g}: G \longrightarrow \mathbb{P}_1$ be the projections. Now $\mathbb{C}_* = \mathbb{C} - \{0\}$ acts holomorphically on $\mathbb{P}_1$ by multiplication. A biholomorphic map α of $M \times \mathbb{P}_1$ onto itself is defined by $\alpha(x,y) = (x,h(x)\, y)$. Then $\alpha(F) = G$. Define $H = h \circ \overleftarrow{f}$. Then $\vec{g} \circ \alpha = H \cdot \vec{f}$. Take $x \in f^{-1}(0)$. Then $(x,0) \in F \cap G$ and $\vec{f}$ is a holomorphic function in a neighborhood of (x,0). Theorem 6.1 implies

$$\nu_g^0(x) = \nu_{\vec{g}}^0(x,0) = \nu_{\vec{g}\circ\alpha}^0(x,0) = \nu_{H\cdot\vec{f}}^0(x,0) = \nu_{\vec{f}}^0(x,0) = \nu_f^0(x).$$

The map $j\colon \mathbb{P}_1 \longrightarrow \mathbb{P}_1$ defined by $j(z) = \frac{1}{z}$ is biholomorphic.

If $x \in f^{-1}(\infty)$, then

$$\nu_g^\infty(x) = \nu_{\vec{g}}^\infty(x,\infty) = \nu_{\vec{g}\circ\alpha}^\infty(x,\infty) = \nu_{H\cdot\vec{f}}^\infty(x,\infty) = \nu_{j(H\cdot\vec{f})}^0(x,\infty)$$

$$= \nu_{j(H)\cdot j(\vec{f})}^0(x,\infty) = \nu_{j(\vec{f})}^\infty(x,\infty) = \nu_f^\infty(x). \qquad \text{q.e.d.}$$

Let E be a holomorphic vector bundle of fiber dimension n over the complex space M of pure dimension m with $q = m - n \geqq 0$. Let s be a holomorphic section in E. If $x \in M - Z(s)$, define $\theta_s(x) = 0$. If $x \in Z(s)$, assume $\dim_x Z(s) = q$. Take a holomorphic frame $(v_1,\ldots,v_n)$ of E over an open neighborhood U of x. Then $s = \sum_{p=1}^{n} f_p v_p$ on U, where $f = (f_1,\ldots,f_n)$ is a holomorphic vector function, q-fibering on a neighborhood of x. Hence $\theta_s(x) = \nu_f(x)$ is defined. By Theorem 6.1 $\theta_s(x)$ does not depend on the choice of the frame and is called the <u>multiplicity of s at x.</u>

If δ is a divisor on M given by a Cousin II distribution $\{f_\lambda, U_\lambda\}_{\lambda\in\Lambda}$, then the <u>multiplicity</u> θ_δ of δ is well defined by $\theta_\delta(x) = \theta_{f_\lambda}(x)$ for each λ with $x \in U_\lambda$. If δ is holomorphic, then δ defines a distinguished line bundle L_δ with a distinguished section s_δ such that $\theta_\delta = \theta_{s_\delta}$.

b) <u>Multiplicities for semi-ample line bundles</u>

Let L be a semi-ample holomorphic line bundle on the complex space N. Let $\eta\colon N \times V \longrightarrow L$ be a semi-amplification with $\dim V = k + 1$. Let M be a complex space of pure dimension m. Take $p \in \mathbb{Z}[0,k]$ with $q = m - p - 1 \geqq 0$. Let $f\colon M \longrightarrow N$ be a holomorphic map. Assume f is adapted to $a \in G_p(V)$ for L at $x \in M$. Then a <u>multiplicity</u> $\theta_f^a[L](x)$ shall be defined. If $x \notin f_a = f^{-1}(E_L[a])$, define $\theta_f^a[L](x) = 0$. If $x \in f_a$, take a holomorphic frame v of L over an open neighborhood U of f(x). Define $\mathfrak{w}\colon U \longrightarrow V^*$ by (3.4). Take $\mathfrak{a}_0,\ldots,\mathfrak{a}_p$ in V such that

$a = \mathbb{P}(\mathfrak{a}_0 \wedge \ldots \wedge \mathfrak{a}_p)$. Define $w_\mu = (\mathfrak{w}; \mathfrak{a}_\mu)$. Let W be an open neighborhood of x such that $f(W) \subseteq U$ and such that $f_a \cap W$ is pure q-dimensional. Then $\mathfrak{v} = (w_0 \circ f, \ldots, w_p \circ f)$ is holomorphic on W with $f_a \cap W = \mathfrak{v}^{-1}(0)$. Hence $\mathfrak{v}$ is q-fibering in a neighborhood of x and the multiplicity $\theta_f^a[L](x) = \nu_{\mathfrak{v}}^0(x)$ is defined.

The definition does not depend on the choices as shall be shown now. Let $\tilde{U}$, $\tilde{v}$, $\tilde{\mathfrak{w}}$, $\tilde{\mathfrak{a}}_\mu$, $\tilde{w}_\mu$, $\tilde{W}$, $\tilde{\mathfrak{v}}$ be another choice. A holomorphic function $g: U \cap \tilde{U} \longrightarrow \mathbb{C} - \{0\}$ exists such that $v = g\tilde{v}$. Hence $\tilde{\mathfrak{w}} = g\,\mathfrak{w}$. Constants $c_{\mu\nu} \in \mathbb{C}$ exist such that $\tilde{\mathfrak{a}}_\mu = \sum_{\nu=0}^{p} c_{\mu\nu}\mathfrak{a}_\nu$. Define the matrix $c = (c_{\mu\nu})$. Then $\det c \neq 0$. On $W \cap \tilde{W}$, we have $\tilde{\mathfrak{v}} = (g \circ f)\, c \cdot \mathfrak{v}$. Theorem 6.1 implies $\nu_{\mathfrak{v}}^0(x) = \nu_{\tilde{\mathfrak{v}}}^0(x)$. Therefore the <u>multiplicity</u> of f to a for L at x is well defined.

If f is adapted to a for L, then $\theta_f^a[L] > 0$ is locally constant on the set of simple points of $f^{-1}(E_L[a])$ with $f^{-1}(E_L[a]) = \operatorname{supp} \theta_f^a[L]$.

The situation becomes more complex if $f: M \rightarrow N$ is meromorphic. Let $\hat{M}$ be the closed graph of f and let $\vec{f}: \hat{M} \longrightarrow N$ and $\overleftarrow{f}: \hat{M} \longrightarrow M$ be the projections. Assume that f is safe and that f is adapted to $a \in G_p(V)$ for L at $x \in M$. Then $\vec{f}$ is adapted to a for L at every $y \in B = \overleftarrow{f}^{-1}(x)$ Define

$$\theta_f^a[L](x) = \sum_{y \in B} \theta_{\vec{f}}^a[L](y) \leqq \infty \tag{6.1}$$

as the multiplicity of f to a for L at x. Since B may be a continuum the sum may become infinite. However, if f is adapted to a for L, then there exists an analytic subset S of $f_a = f^{-1}(E_L[a])$ with $\dim S < q$ and such that $B = B(x)$ is finite for all $x \in f_a - S$. Hence $\theta_f^a[L](x) < \infty$ if $x \in f_a - S$. Clearly $\theta_f^a[L] > 0$ on f_a and $\theta_f^a[L] = 0$ on $S - f_a$. For technical reasons, we define $\theta_f^a[L](x) = 0$ for all $x \in f_a$ for which the sum in (6.1) is infinite.

Let χ be a continuous form of bidegree (q,q) on M. Let A be an analytic subset of M and let A_q be the union of all q-dimensional branches of A. Let $j: A_q \longrightarrow M$ be the inclusion. The pull back $j^*(\chi)$ exists (Tung [87]). Then χ is said to be integrable over A if $j^*(\chi)$ is integrable over A_q. Define

$$(6.2) \qquad \int_A \chi = \int_{A_q} \chi = \int_{A_q} j^*(\chi).$$

Lemma 6.3. Assume the meromorphic map $f: M \longrightarrow N$ is adapted to $a \in G_p(V)$ for L. Abbreviate $A = f^{-1}(E_L[a])$ and $\hat{A} = \overleftrightarrow{f}^{-1}(E_L[a])$. Let χ be a continuous form of bidegree (q,q) on M and define $\hat{\chi} = \overleftarrow{f}^*(\chi)$ on $\hat{M}$. Assume that $\theta_{\overrightarrow{f}}^a [L]\, \hat{\chi}$ is integrable over $\hat{A}$. Then $\theta_f^a\, [L]\chi$ is integrable over A and

$$(6.3) \qquad \int_A \theta_f^a\, [L]\, \chi = \int_{\hat{A}} \theta_{\overrightarrow{f}}^a\, [L]\, \hat{\chi}$$

Proof. The restriction $\rho = \overleftarrow{f}: \hat{A} \longrightarrow A$ is proper and surjective. Assume $\hat{A} \neq \emptyset$. Then $\hat{A}$ is pure q-dimensional and $\dim A \leqq q$. The sets $D = \{x \in \hat{A} \mid \operatorname{rank}_x \rho < q\}$ and $S = \rho(D)$ are analytic with $\dim S < q$. Define $\hat{S} = \rho^{-1}(S)$. Let $j: S \longrightarrow M$ and $\hat{j}: \hat{S} \longrightarrow \hat{M}$ be the inclusions. Let $\rho_0 = \rho: \hat{S} \longrightarrow S$ be the restriction. Then $j \circ \rho_0 = \overleftarrow{f} \circ \hat{j}$. Here $\dim S < q$ implies $j^*(\chi) = 0$. Hence $\hat{j}^*(\hat{\chi}) = \rho_0^*(j^*(\chi)) = 0$. Now $\hat{F} = \hat{A} - \hat{S}$ and $F = A - S$ are pure q-dimensional and $\rho: \hat{F} \longrightarrow F$ is surjective, light and proper. Write $g = \theta_{\overrightarrow{f}}^a\, [L]$ and $h = \theta_f^a\, [L]$. Then (6.1) and 6.2) imply

$$\int_{\hat{A}} g\, \hat{\chi} = \int_{\hat{F}} g\, \hat{\chi} = \int_F h\, \chi = \int_A h\, \chi \qquad \text{q.e.d.}$$

If $K = A \cap \operatorname{supp} \chi$ is compact, then $\rho^{-1}(K)$ is compact and (6.3) holds. If $\chi \geqq 0$ and $\theta_f^a\, [L]\, \chi$ is integrable over A, then $\theta_{\overrightarrow{f}}^a\, [L]\, \hat{\chi}$ is integrable over $\hat{A}$ and (6.3) holds.

III. The First Main Theorem

7. Integral Theorems

a) Remarks on differential forms

Bloom and Herrera [3] introduced differential forms on complex spaces. A detailed account with complete proofs is given by Tung [87]. For short surveys see King [37] and Stoll [80]. Here some basic facts shall be recalled and some additional remarks shall be made.

Let M be a complex space. Let $\mathfrak{R}(M)$ be the set of simple points. A patch $\alpha: U_\alpha \longrightarrow U'_\alpha$ is a biholomorphic map of an open subset U_α of M onto an analytic subset U'_α of an open subset G_α of $\mathbb{C}^{k_\alpha}$. Let $j_\alpha: U'_\alpha \longrightarrow G_\alpha$ be the inclusion and define $\alpha_0 = j_\alpha \circ \alpha \mid \mathfrak{R}(U_\alpha)$. Let $\mathfrak{P}(M)$ be the set of patches of M. The patch is smooth if $U'_a = G_\alpha$. Let $\mathfrak{S}(M)$ be the set of smooth patches of M. A form χ of class C^k on $\mathfrak{R}(M)$ is called of class C^k on M if for each $a \in M$ there exists $\alpha \in \mathfrak{P}(M)$ with $a \in U_\alpha$ and a form $\hat{\chi}$ of class C^k on G_α (called an extension of χ) such that $\alpha_0^*(\hat{\chi}) = \chi \mid \mathfrak{R}(U_\alpha)$. The usual elementary properties of differential forms hold for forms on M, especially the pull back under holomorphic and even differential maps hold. The form χ of bidegree (p,p) and class C^k on M is said to be non-negative ($\chi \geqq 0$) if for every holomorphic map $\varphi: G \longrightarrow M$ of an open subset $G \neq \emptyset$ of C^p the pull back $\varphi^*(\chi)$ is non-negative. If M is a manifold, then χ is said to be positive ($\chi > 0$) if for each such smooth holomorphic map $\varphi: G \longrightarrow M$ the pull back $\varphi^*(\chi)$ is positive. If M is a complex space, then χ is said to be positive ($\chi > 0$), if for each $a \in M$ a patch $\alpha \in \mathfrak{P}(M)$ and a positive extension $\hat{\chi}$ of χ on G_α exist. If M is pure m-dimensional, positive forms of class C^∞ and bidegree (p,p) exist for each $p \in \mathbb{Z}[0,m]$. If $\chi > 0$ has bidegree (p,p) and if $\psi > 0$ has bidegree (1,1), then $\chi \wedge \psi > 0$. If $\psi = \overline{\psi}$ and $\chi > 0$ are forms of class C^k and bidegree (p,p), and if K is compact, then a constant $c > 0$ exist such that $c\chi + \psi > 0$ on an open neighborhood U of K.

Take $\emptyset \neq S \subseteq M$. Let χ be a form of class C^k on M. Then $\mathrm{supp}_S\ \chi \subseteq S$ and $a \in S - \mathrm{supp}_S\ \chi$ if and only if a patch $\alpha \in \mathfrak{P}(M)$ with $a \in U_\alpha$ and an extension $\hat{\chi}$ of χ for α exist such that $\hat{\chi} \mid \alpha(S \cap U_\alpha) = 0$. Abbreviate $\mathrm{supp}\ \chi = \mathrm{supp}_M \chi$. Let $\mathfrak{H}^p$ be the p-dimensional Hausdorff measure on $\mathbb{R}^m$ (Federer [21] p 171). Then S is said to have locally finite $\mathfrak{H}^p$-measure (respectively zero $\mathfrak{H}^p$-measure) if for each $a \in M$ there exists $\alpha \in \mathfrak{P}(M)$ with $a \in U_\alpha$ such that $\mathfrak{H}^p(\alpha(U_\alpha \cap S)) < \infty$ (respectively $\mathfrak{H}^p(\alpha(U_\alpha \cap S)) = 0$.)

Let G be an open subset of a pure m-dimensional complex space M. Then $\mathfrak{R}(G)$ has a largest boundary manifold dG of class C^∞ in $\mathfrak{R}(M)$ (of course dG could be empty). Now G is called a Stokes domain, if $\overline{G}$ is compact, if ∂G has locally finite $\mathfrak{H}^{2m-1}$-measure and if $\partial G - dG$ has zero $\mathfrak{H}^{2m-1}$-measure. If so and if χ is a form of class C^1 and degree $2m - 1$ on M, then Stokes Theorem holds.

$$\int_G d\chi = \int_{dG} \chi \, . \tag{7.1}$$

Here dG is oriented to the exterior of $\mathfrak{R}(G)$. The assumptions are too strong. Let $\Sigma(M)$ be the set of singular points of M. Define $K = \overline{G} \cap \mathrm{supp}\ \chi$ and $T = \mathrm{supp}_{\partial G}\ \chi$. Then G is said to be Stokes admissible for χ if there exists a thin analytic set $E \supseteq \Sigma(M)$ such that $K \cap (\partial G -$ has locally finite $\mathfrak{H}^{2m-1}$-measure on $\mathfrak{R}(M)$, if $T - (E \cup dG)$ has zero $\mathfrak{H}^{2m-1}$-measure on $\mathfrak{R}(M)$ and if K is compact. For this definition χ may have any degree p. If $p = 0$ and $\chi = 1$, then G is said to be a weak Stokes domain if G is Stokes admissible for 1. (Then $K = \overline{G}$ and $T = \partial G$.)

If G is Stokes admissible for a form χ of class C^1 and degree $(2m-1)$ and if χ is integrable over dG, then (7.1) holds. The integrability of χ over dG can be replaced by the following condition: Let $j: dG \longrightarrow \mathfrak{R}(M)$ be the inclusion. Assume χ is real. Define

$$\Gamma_+ = \{x \in dG \mid j^*(\chi) > 0\} \qquad \Gamma_- = \{x \in dG \mid j^*(\chi) < 0\}.$$

Then χ is said to have <u>separate</u> signs on ∂G if $\overline{\Gamma}_+ \cap \overline{\Gamma}_- = \emptyset$. If so and if χ is of class C^1 and if G is Stokes admissible for χ, then $j^*(\chi)$ is integrable over dG and (7.1) holds.

Lemma 7.1. Let $f: M \longrightarrow N$ be a proper, holomorphic map of strict rank m between complex spaces of pure dimension m. Let χ be a continuous form of degree p on N. Let G be Stokes admissible for χ. Then $f^{-1}(G)$ is Stokes admissible for $f^*(\chi)$.

For a proof see [82]. In [67] Satz 4.5 and [79] Lemma 3.1 the following useful fact was established.

Lemma 7.2. Let G be open in the complex space M of pure dimension m. Take $a \in dG$. Let $j: dG \longrightarrow \mathfrak{R}(M)$ be the inclusion. Let $\psi: M \longrightarrow \mathbb{R}$ be of class C^1. Assume that a neighborhood U of a exists such that either $\psi(x) \geqq \psi(a)$ for all $x \in U - G$ or $\psi(x) \geqq \psi(a)$ for all $x \in U \cap G$. As always dG is oriented to the exterior of G. Let $\chi \geqq 0$ be a continuous form of bidegree (m-1,m-1) on M. Then $j^*(d^c\psi \wedge \chi)(a) \geqq 0$. If $\chi(a) > 0$ and $(d\psi)(a) \neq 0$, then $j^*(d^c\psi \wedge \chi)(a) > 0$.

Lemma 7.3. Let M and N be complex spaces of pure dimensions m and n respectively. Let χ and ψ be continuous forms of degree q on M respectively of degree p on N with $p + q = 2m$. Let $f: M \longrightarrow N$ be a meromorphic map. Then $f^*(\psi) \wedge \chi$ is locally integrable over M.

Proof. If the form is lifted into the graph of M, the form becomes continuous hence locally integrable. However the graph of M is a modification of M and does not disturb integrability; q.e.d. (For a detailed proof see [82] Lemma 10.1)

b) The Jensen formula

Let G be open in the complex space M. Then $\psi: M \longrightarrow \mathbb{R}$ is said to have a <u>consistent sign</u> around ∂G if $\psi \mid \partial G = 0$ and if an open neighborhood W of ∂G exists such that either $\psi \mid (G \cap W) \geqq 0$ or $\psi \mid (G \cap W) \leqq 0$ or $\psi \mid (W - G) \geqq 0$ or $\psi \mid (W - G) \leqq 0$.

Theorem 7.4. (Jensen formula). Let χ be a form of class C^2 and of bidegree (m-1, m-1) on the complex space M of pure dimension m. Let G be Stokes admissible for χ. Assume the function $\psi: M \longrightarrow \mathbb{R}$ of class C^2

has consistent sign around ∂G. Let f be a meromorphic function on M. Assume $A = f^{-1}(0) \cup f^{-1}(\infty)$ is thin in M. Let θ_f be the divisor multiplicity of f. Then

$$(7.2) \qquad \int_G \log |f| \, dd^c(\psi\chi) = \tfrac{1}{2} \int_{A \cap G} \theta_f \psi\chi + \int_{dG} \log |f| \, d^c\psi \wedge \chi.$$

<u>Proof.</u> Define $N = f^{-1}(0)$ and $P = f^{-1}(\infty)$. At first assume that $f: M \longrightarrow \mathbb{P}_1$ is holomorphic. Then $N \cap P = \emptyset$. If $P \cap \operatorname{supp} \chi = \emptyset$, Theorem A.8 with $n = 1$ and $s = 0$ implies (7.2). If $N \cap \operatorname{supp} \chi = \emptyset$, set $g = 1/f$. Then $\theta_g = -\theta_f$ and $N = g^{-1}(\infty)$. Hence (7.2) holds. Assume $P \cap \operatorname{supp} \chi \neq \emptyset \neq N \cap \operatorname{supp} \chi$. Take a function $\lambda: M \longrightarrow \mathbb{R}[0,1]$ of class C^∞ such that $\lambda = 0$ in a neighborhood of P. Then (7.2) holds for $\lambda\chi$ and for $(1 - \lambda)\chi$ hence for χ. If $f: M \longrightarrow \mathbb{P}_1$ is meromorphic, let $\hat{M}$ be the graph and apply (7.2) to the projection $\vec{f}: \hat{M} \longrightarrow \mathbb{P}_1$. Since the projection $\overleftarrow{f}: \hat{M} \to M$ is a modification, (7.2) follows for f; q.e.d.
(For a detailed proof see [82] Theorem 9.12).

If χ has compact support, Lelong's theorem

$$\int_A \theta_f \, \chi = \int_M \log |f|^2 \, dd^c\chi$$

follows, which says that the current $dd^c[\log |f|^2]$ equals integration over θ_f. If M is compact and $d\chi = 0$, Wirtinger's theorem follows

$$\int_A \theta_f \, \chi = 0.$$

Also the Jensen formulas of [67] Satz 6.5 and of [80] Theorem 5.2 follow easily.

c) <u>The Residue Theorem for line bundles</u>

For this section compare Griffiths-King [33] 1(c). The following <u>General Assumptions</u> shall be made.

(A1) Let M be a complex space of pure dimension $m > 0$.

(A2) Let N be a complex space.

(A3) Let L be a holomorphic line bundle over N with a hermitian metric κ along the fibers of L.

(A4) Let V be a complex vector space of dimension $k + 1 \geqq 1$ with a hermitian metric ℓ.

(A5) Let η: $N \times V \longrightarrow L$ be a semi-amplification of L.

(A6) Let f: $M \longrightarrow N$ be a meromorphic map.

Theorem 7.5. Assume (A1) - (A6). Take $a \in G_p(V)$ with $0 \leqq p \leqq k$ and $q = m - p - 1 \geqq 0$. Assume that f is adapted to a for L. Let χ be a form of class C^2 and bidegree (q,q) on M. Let G be Stokes admissible for χ. Let ψ: $M \longrightarrow \mathbb{R}$ be a function of class C^2 which has consistent sign around ∂G. Abbreviate $A = G \cap f^{-1}(E_L[a])$. Then

$$\int_G \psi\, f^*(c(L,\kappa)^{p+1}) \wedge \chi = \int_A \theta_f^a[L]\ \psi\chi + \int_G f^*(\Lambda_L[a]_\kappa) \wedge dd^c(\psi\chi)$$

(7.3)

$$-\int_{dG} f^*(\Lambda_L[a]_\kappa) \wedge d^c\psi \wedge \chi$$

(7.4)
$$\int_G f^*(\hat{\Lambda}_L[a]_\kappa) \wedge dd^c(\psi\chi) = \int_{dG} f^*(\hat{\Lambda}_L[a]_\kappa \wedge d^c\psi \wedge \chi.$$

Proof. At first assume that f is holomorphic. Using a partition of unity on the compact set $\overline{G} \cap \operatorname{supp} \chi$, it suffices to construct to each $z \in M$ an open neighborhood W such that (7.3) and (7.4) hold if χ has compact support which is contained in W.

Define $f_a = f^{-1}(E_L[a])$. Take a holomorphic frame v of L over an open neighborhood U of f(z). Take an open neighborhood W_0 of z such that $f(W_0) \subseteq U$ and such that $W_0 \cap f_a$ is either empty or pure q-dimensional. Define $\mathfrak{w}$: $U \longrightarrow V^*$ by (3.4). Let $\mathfrak{a}_0, \ldots, \mathfrak{a}_k$ be an orthonormal base of V such that $\mathbb{P}(\mathfrak{a}_0 \wedge \cdots \wedge \mathfrak{a}_p) = a$. Let $\alpha_0, \ldots, \alpha_k$ be the dual base. Then $\mathfrak{w} = w_0\alpha_0 + \cdots + w_k\alpha_k$.

A holomorphic map $\mathfrak{v} = (w_0 \circ f, \ldots, w_p \circ f)$ of W_0 into $\mathbb{C}^{p+1}$ is defined with $\mathfrak{v}^{-1}(0) = W_0 \cap f_a$. If $W_0 \cap f_a = \emptyset$, take $W = W_0$. If $W_0 \cap f_a \neq \emptyset$, then $\mathfrak{v}^{-1}(0)$ is pure q-dimensional and an open neighborhood W of a in W_0 is taken such that $\mathfrak{v}|W$ is q-fibering. Then $\|a,f\|_\kappa = |\mathfrak{v}|\,|v \circ f|_\kappa$ on W and $c(L,\kappa) = -\, dd^c \log |v|_\kappa^2$ on U. Hence

$$f^*(\Phi_L[a]) = dd^c \log \|a,f\|_\kappa^2 + f^*(c(L,\kappa)) = dd^c \log |\mathfrak{v}|^2 = \mathfrak{v}^*(\omega).$$

Define $h = -\log |v \circ f|_\kappa^2$. Then $dd^c h = f^*(c(L,\kappa))$ and

$$f^*(\Lambda_L[a]_\kappa) = (h - \log |\mathfrak{v}|^2) \sum_{\mu=0}^{p} \mathfrak{v}^*(\omega^\mu) \wedge (dd^c h)^{p-\mu}.$$

By definition $\theta_f^a[L] = \nu_{\mathfrak{v}}^0 = \nu_{\mathfrak{v}}$ on $f_a \cap W$.

Assume that χ has compact support which is contained in W. Then $G_0 = G \cap W$ is Stokes admissible for χ. Take $\mu \in \mathbb{Z}[0,p]$. Theorem A4 (Appendix) and its Remark imply

$$\int_{G_0} d(\psi\, \chi\, \mathfrak{v}^*(\omega^\mu) \wedge (dd^c h)^{p-\mu} \wedge d^c h) = 0$$

$$\int_{G_0} h\ \mathfrak{v}^*(\omega^\mu) \wedge (dd^c h)^{p-\mu} \wedge dd^c(\psi\chi)$$

$$= \int_{dG_0} h\, \mathfrak{v}^*(\omega^\mu) \wedge (dd^c h)^{p-\mu} \wedge d^c(\psi\chi)$$

$$- \int_{G_0} dh \wedge \mathfrak{v}^*(\omega^\mu) \wedge (dd^c h)^{p-\mu} \wedge d^c(\psi\chi)$$

$$= \int_{dG_0} h\, \mathfrak{v}^*(\omega^\mu) \wedge (dd^c h)^{p-\mu} \wedge d^c \psi \wedge \chi$$

$$+ \int_{G_0} \psi\, \mathfrak{v}^*(\omega^\mu) \wedge (dd^c h)^{p-\mu+1} \wedge \chi$$

where (3) in the Appendix was used. If $\mu \in \mathbb{Z}[0, p-1]$, Theorem A7 implies

$$\int_{G_0} (\log |\mathfrak{v}|^2)\ \mathfrak{v}^*(\omega^\mu) \wedge (dd^c h)^{p-\mu} \wedge dd^c(\psi\chi)$$

$$= \int_{G_0} \psi\ \mathfrak{v}^*(\omega^{\mu+1}) \wedge (dd^c h)^{p-\mu} \wedge \chi$$

$$+ \int_{dG_0} (\log |\mathfrak{v}|^2)\ \mathfrak{v}^*(\omega^\mu) \wedge (dd^c h)^{p-\mu} \wedge d^c\psi \wedge \chi.$$

If $\mu = p$, Theorem A7 implies

$$\int_{G_0} (\log |\mathfrak{v}|^2)\ \mathfrak{v}^*(\omega^p) \wedge dd^c(\psi\chi)$$

$$= \int_{G_0 \cap \mathfrak{f}_a} \nu_{\mathfrak{v}} \psi\ \chi + \int_{dG_0} (\log |\mathfrak{v}|^2)\ \mathfrak{v}^*(\omega^p) \wedge d^c\psi \wedge \chi.$$

Since supp $\chi \subset W$, addition implies

$$\int_G f^*(\Lambda_L[a]_\kappa) \wedge dd^c(\psi\chi) - \int_{dG} f^*(\Lambda_L[a]_\kappa) \wedge d^c\psi \wedge \chi$$

$$= \sum_{\mu=0}^{p} \int_{G_0} \psi\ \mathfrak{v}^*(\omega^\mu) \wedge (dd^c h)^{p-\mu+1} \wedge \chi$$

$$- \sum_{\mu=0}^{p-1} \int_{G_0} \psi\ \mathfrak{v}^*(\omega^{\mu+1}) \wedge (dd^c h)^{p-\mu} \wedge \chi - \int_{G_0 \cap \mathfrak{f}_a} \nu_{\mathfrak{v}} \psi\ \chi$$

$$= \int_G \psi\ f^*(c(L,\kappa)^{p+1}) \wedge \chi - \int_A \theta_f^a[L]\ \psi\ \chi$$

which proves (7.3) in this case.

If $\mu \in \mathbb{Z}[0,p]$, Theorem A7 shows

$$\int_{G_0} \mathfrak{v}^*(\omega^\mu) \wedge (dd^c h)^{p-\mu} \wedge dd^c(\psi\chi)$$

$$= \int_{dG_0} \mathfrak{v}^*(\omega^\mu) \wedge (dd^c h)^{p-\mu} \wedge d^c\psi \wedge \chi.$$

Addition yields (7.4) in this case.

If f is meromorphic, let $\hat{M}$ be the graph of f. Then the theorem holds for the projection $\vec{f}: \hat{M} \longrightarrow N$. Since the projection $\overleftarrow{f}: \hat{M} \longrightarrow M$ is a modification (7.3) and (7.4) follow. Here Lemma 7.1 and Lemma 6.3 are used; q.e.d. (For more details see [82] Theorem 9.16).

Recall that the hermitian metric ℓ on V defines a hermitian metric ℓ along the fibers of $L|N_\infty$ where $N_\infty = N - E_L[\infty]$. Let φ be the dual classification map of η and let $f: M \longrightarrow N$ be a safe holomorphic map. Using (4.15)-(4.21) Theorem 7.5 can be applied to the meromorphic map $\varphi_f = \varphi \circ f$ and to the hyperplane section bundle over $\mathbb{P}(V^*)$ to obtain Theorem 7.5 for ℓ. If f is safe of order p + 1 and if f is strictly adapted to a for L, then

$$\theta^a_{\varphi_f}[S_0(V^*)^*] = \theta^a_f[L]$$

on $f^{-1}(E_L[a] \cap M_\infty$ whose complement is thin in $f^{-1}(E_L[a])$. After Theorem 7.5 for ℓ is established for holomorphic maps, it can be extended to meromorphic maps by going over the graph. For more details see [82] Theorem 9.17 and 9.18. Hence the following result is obtained:

<u>Theorem 7.6</u>. Assume (A1)-(A6). Take $a \in G_p(V)$ with $0 \leqq p \leqq k$ and $q = m - p - 1 \geqq 0$. Assume that f is strictly adapted to a for L. Assume that f is safe of order p + 1 for L. Let χ be a form of class C^2 and bidegree (q,q) on M. Let G be Stokes admissible for χ. Let $\psi: M \longrightarrow \mathbb{R}$ be a function of class C^2 which has consistent sign around ∂G.

Abbreviate $A = G \cap f^{-1}(E_L[a])$. Then

$$\int_G \psi\, f^*(c(L,\ell)^{p+1} \wedge \chi = \int_A \theta_f^a[L]\, \psi\chi + \int_G f^*(\Lambda_L[a] \wedge dd^c(\psi\chi)$$

$$(7.5) \qquad - \int_{dG} f^*(\Lambda_L[a]) \wedge d^c\psi \wedge \chi$$

$$(7.6) \qquad \int_G f^*(\hat{\Lambda}_L[a]) \wedge dd^c(\psi\chi) = \int_{dG} f^*(\hat{\Lambda}_L[a] \wedge d^c\psi \wedge \chi.$$

If K is compact in M, an open neighborhood G of K exists such that G is a Stokes domain. Also a function ψ of class C^∞ and compact support in G exists on M such that $\psi = 1$ in a neighborhood of K. Hence $f^*(\Lambda_L[a]_\kappa) \wedge \chi$ respectively $f^*(\Lambda_L[a]) \wedge \chi$ are locally integrable over M under the assumptions of Theorem 7.5 respectively Theorem 7.6.

Compare the results of this section with Griffiths-King [33] 1(c).

8. The general First Main Theorem

Let G and g be weak Stokes domains in M with $\overline{g} \subset G$. Then (G,g,ψ) is said to be a condensor in M if $\psi: M \longrightarrow \mathbb{R}$ is a continuous function such that $\psi|\overline{g} = R$ is constant, such that $0 \leqq \psi \leqq R$ on M such that $\psi|\partial G = 0$ and such that $\psi|(\overline{G}-g)$ extends to a function of class C^2 on M. Then $\overline{g}$ and $\overline{G}$ are compact and ψ(respectively $\psi-R$) has consistent sign around ∂G(respectively ∂g).

In addition to (A1)-(A6) assume

(A7) Let (G,g,ψ) be a condensor on M.

(A8) Take $p \in \mathbb{Z}[0,k]$ with $q = m - p$.

(A9) Let $\chi \geqq 0$ be a non-negative form of bidegree (q,q) and class C^2 on M such that $d\chi = 0$.

Then the <u>spherical image</u> A_p and the <u>characteristic</u> T_p of order p of f for (L,κ) exist by Lemma 7.3 and are defined by

$$(8.1) \qquad A_p(L,\kappa,f) = \int_G f^*(c(L,\kappa)^p) \wedge \chi$$

$$(8.2) \qquad T_p(L,\kappa,f) = \int_G \psi\, f^*(c(L,\kappa)^p) \wedge \chi .$$

If $c(L,\kappa) \geqq 0$, then $A_p(L,\kappa,f) \geqq 0$ and $T_p(L,\kappa,f) \geqq 0$. If f is safe, then also the <u>spherical image</u> $A_p(L,f)$ and the <u>characteristic</u> $T_p(L,f)$ of order p of f for L and the <u>average deficit</u> $\Delta_p(L,f)$ exist

$$(8.3) \qquad A_p(L,f) = \int_G f^*(c(L,\ell)^p) \wedge \chi \geqq 0$$

$$(8.4) \qquad T_p(L,f) = \int_G \psi\, f^*(c(L,\ell)^p \wedge \chi \geqq 0$$

$$(8.5) \qquad \Delta_p(L,f) = \int_G f^*(c(L,\ell)^{p-1}) \wedge d^c d\psi \wedge \chi .$$

In order to prove the existence, let $\hat{M}$ be the graph of f and let $\overleftarrow{f}\colon \hat{M} \longrightarrow M$ and $\overrightarrow{f}\colon \hat{M} \longrightarrow N$ be the projections. Let φ be the dual classification map of η. Then $\overrightarrow{f}$ is safe and $\varphi \circ \overrightarrow{f}^*$ is meromorphic. Since $\overrightarrow{f}^*(c(L,\ell))) = (\varphi\circ\overrightarrow{f})^*(\omega)$, the form $\overrightarrow{f}^*(c(L,\ell)) \wedge \overleftarrow{f}^*(\chi)$ is locally integrable over $\hat{M}$. Since $\overleftarrow{f}$ is a modification, the integrals exist. For more details see [82] Lemma 11.1.

Observe that $\Delta_1 = \Delta(L,f)$ does not depend on L and f.

If f is adapted to $a \in G_{p-1}(V)$ for L, put $A = G \cap f^{-1}(E_L[a])$ and define the <u>counting function</u> and the <u>valence function</u> of f or L to a respectively by

(8.6) $$n_a(L,f) = \int_A \theta_f^a[L]\, \chi \geqq 0$$

(8.7) $$N_a(L,f) = \int_A \theta_f^a[L]\, \psi\, \chi \geqq 0.$$

The integrals exist by Lemma 6.3. Also the compensation functions $m_a(L,\kappa,f)$ and $m_a^0(L,\kappa,f)$ and the deficit $D_a(L,\kappa,f)$ of f for (L,κ) to a exist by Theorem 7.5:

(8.8) $$m_a(L,\kappa,f) = \int_{dG} f^*(\Lambda_L[a]_\kappa) \wedge (-d^c\psi) \wedge \chi$$

(8.9) $$m_a^0(L,\kappa,f) = \int_{dG} f^*(\Lambda_L[a]_\kappa) \wedge (-d^c\psi) \wedge \chi$$

(8.10) $$D_a(L,\kappa,f) = \int_G f^*(\Lambda_L[a]_\kappa) \wedge d^c d\psi \wedge \chi.$$

If $c(L,\kappa) \geqq 0$ and if κ is distinguished, then $m_a(L,\kappa,f) \geqq 0$ and $m_a^0(L,\kappa,f) \geqq 0$ by Lemma 7.2. In (8.10), the integration goes over $G - \overline{g}$ in fact.

If f is safe of order p for L and if f is strictly adapted to a for L then the compensation functions $m_a(L,f)$ and $m_a^0(L,f)$ and the deficit $D_a(L,f)$ of f for L to a exist by Theorem 7.6:

(8.11) $$m_a(L,f) = \int_{dG} f^*(\Lambda_L[a] \wedge (-d^c\psi) \wedge \chi \geqq 0$$

(8.12) $$m_a^0(L,f) = \int_{dg} f^*(\Lambda_L[a]) \wedge (-d^c\psi) \wedge \chi \geqq 0$$

(8.13) $$D_a(L,f) = \int_G f^*(\Lambda_L[a]) \wedge (d^c d\psi) \wedge \chi.$$

If f is safe of order p for L, the <u>average compensation</u> functions

(8.14) $$\mu_p(L,f) = - \int_{dG} f^*(c(L,\ell)^{p-1}) \wedge d^c\psi \wedge \chi \geqq 0$$

(8.15) $$\mu_p^0(L,f) = - \int_{dG} f^*(c(L,\ell)^{p-1}) \wedge d^c\psi \wedge \chi \geqq 0$$

exist as consideration of the meromorphic map $\varphi\circ\vec{f}$ and the modification $\overleftarrow{f}$ show. Observe that $\mu_1 = \mu_1(L,f)$ and $\mu_1^0 = \mu_1^0(L,f)$ do not depend on L and f.

If p = 1, the index is also dropped:

$A(L,\kappa,f) = A_1(L,\kappa,f)$ $\qquad A(L,f) = A_1(L,f)$

$T(L,\kappa,f) = T_1(L,\kappa,f)$ $\qquad T(L,f) = T_1(L,f)$

$\Delta(L,f) = \Delta_1(L,f) = \Delta$

$\mu(1,f) = \mu_1(L,f) = \mu_1$ $\qquad \mu^0(L,f) = \mu_1^0(L,f) = \mu_1^0.$

<u>Theorem 8.1</u>. <u>First Main Theorem</u>. Assume (A1) - (A9). Take $a \in G_{p-1}(V)$. Let f be adapted to a for L, then

(8.16) $$T_p(L,\kappa,f) = N_a(L,f) + m_a(L,\kappa,f) - m_a^0(L,\kappa,f) - D_a(L,\kappa,f).$$

If f is safe of order p for L and if f is strictly adapted to a for L, then

(8.17) $$T_p(L,f) = N_a(L,f) + m_a(L,f) - m_a^0(L,f) - D_a(L,f).$$

<u>Proof.</u> Let $j: dG \longrightarrow \mathbb{R}(M)$ and $j_0: dg \longrightarrow \mathbb{R}(M)$ be the inclusions.

A function $\tilde{\psi}: M \longrightarrow \mathbb{R}$ of class C^2 exists such that $\tilde{\psi}|(\overline{G} - g) = \psi$. Then $j^*(d^c\tilde{\psi}) = j^*(d^c\psi)$ and $j_0^*(d^c\tilde{\psi}) = j_0^*(d^c\psi)$. Define $A = f^{-1}(E_L[a]) \cap G$. Theorem 7.5 implies

$$\int_G \tilde{\psi}\, f^*(c(L,\kappa)^p) \wedge \chi = \int_A \theta_f^a[L]\, \tilde{\psi}\chi - \int_G f^*(\Lambda_L[a]_\kappa) \wedge d^c d\tilde{\psi} \wedge \chi$$
$$- \int_{dG} f^*(\Lambda_L[a]_\kappa) \wedge d^c\psi \wedge \chi$$

$$\int_g (R - \tilde{\psi})\, f^*(c(L,\kappa)^p) \wedge \chi = \int_{A\cap g} \theta_p^a[L]\, (R - \tilde{\psi})\, \chi$$
$$+ \int_g f^*(\Lambda_L[a]_\kappa) \wedge d^c d\tilde{\psi} \wedge \chi$$
$$+ \int_{dG} f^*(\Lambda_L[a]_\kappa) \wedge d^c\psi \wedge \chi.$$

Since $\tilde{\psi}|(\overline{G} - g) = \psi$ and $\tilde{\psi} + (R - \tilde{\psi}) = R = \psi$ on g, addition yields (8.16). Using Theorem 7.6, the identity (8.16) is obtained the same way; q.e.d.

The dependence of these value distribution functions on their parameters is of interest, but is left to the reader. Compare the results of this section with Griffiths-King [33] 5(b) and with Stoll [74]. The basic structure of the proof of the First Main Theorem is already contained in Weyl [89] and Stoll [67].

9. Integral Average

In this section, I denotes the integral average operator over $G_{p-1}(V)$. Observe the shift in the index. Define

(9.1)
$$c(p,k) = \sum_{\nu=1}^{p} \sum_{\mu=0}^{k-p} \frac{1}{\nu+\mu}.$$

Lemma 9.1. Assume (A1) - (A6) and (A8). Let K be a measurable subset of M such that $\overline{K}$ is compact. Let f be safe of order $p \geqq 1$. Let χ be a continuous form of bidegree (q+1, q+1) on M. Then

(9.2) $$I(\int_K f^*(\Lambda_L[a]) \wedge \chi) = c(p,k) \int_K f^*(c(L,\ell)^{p-1}) \wedge \chi$$

(9.3) $$I(\int_K f^*(\hat{\Lambda}_L[a]) \wedge \chi) = p \int_K f^*(c(L,\ell)^{p-1}) \wedge \chi.$$

Proof. Let $\mathfrak{A}^s_{p-1}(M,L,f)$ be the set of all $a \in G_{p-1}(V)$ such that f is strictly adapted to a for L. At first assume that f is holomorphic. Then $G_{p-1}(V) - \mathfrak{A}^s_{p-1}(M,L,f)$ has measure zero and

$$\int_K f^*(\Lambda_L[a]) \wedge \chi$$

exists for all $a \in \mathfrak{A}^s_{p-1}(M,L,f)$ by Theorem 7.6 and the subsequent remark. Let $\varphi: N \longrightarrow \mathbb{P}(V^*)$ be the dual classification map. Then $\varphi \circ f = \varphi_f: M \longrightarrow \mathbb{P}(V^*)$ is meromorphic and

$$f^*(c(L,\ell)^{p-1}) \wedge \chi = \varphi_f^*(\ddot{\omega}^{p-1}) \wedge \chi$$

is integrable over K by Lemma 7.3.

Assume $\chi \geqq 0$. Then $f^*(\Lambda_L[0] \wedge \chi) \geqq 0$. The operator I commutes with pull backs. Hence (4.20), (4.17), Theorem 4.9 and Fubini's theorem imply (9.2), and (9.3). If χ is real, take a continuous form $\zeta > 0$ of bidegree (q+1,q+1) such that $\chi + \zeta > 0$ in an open neighborhood of K. Then (9.2) and (9.3) holds for $\chi + \zeta$ and for ζ and by subtraction for χ. If χ is complex valued split χ into real and imaginary parts.

If f is meromorphic, let $\hat{M}$ be the graph of f and let $\vec{f}: \hat{M} \longrightarrow N$ and $\overleftarrow{f}: \hat{M} \longrightarrow M$ be the projections. Then $\vec{f}$ is safe of order p for L. Therefore (9.2) and (9.3) hold for $\overleftarrow{f}^{-1}(K), \vec{f}$ and $\overleftarrow{f}^*\chi$. Since $\overleftarrow{f}$ is a modification, (9.2), and (9.3) follow for K, f, and χ; q.e.d.

If $a \in G_{p-1}(V)$, define $f_a = f^{-1}(E_L[a])$.

Theorem 9.2. Assume (A1) - (A6) and (A8). Let ζ be a form of class C^2 and bidegree (q,q) on M. Let G be Stokes admissible for ζ. Let $\psi: M \longrightarrow \mathbb{R}$ be a function of class C^2 which has consistent sign around ∂G. Let f be safe of order $p \geqq 1$. Then

(9.4) $$I\left(\int_{dG} f^*(\Lambda_L[a]) \wedge d^c\psi \wedge \zeta\right) = c(p,k) \int_{dG} f^*(c(L,\ell)^{p-1}) \wedge d^c\psi \wedge \zeta$$

(9.5) $$I\left(\int_{dG} f^*(\hat{\Lambda}_L[a]) \wedge d^c\psi \wedge \zeta\right) = p \int_{dG} f^*(c(L,\ell)^{p-1}) \wedge d^c\psi \wedge \zeta$$

(9.6) $$I\left(\int_{G\cap f_a} \psi \, \theta_f^a[L]\zeta\right) = \int_G \psi \, f^*(c(L,\ell)^p) \wedge \zeta .$$

Proof. By Theorem 7.6, the integrals

$$g(a,\zeta) = \int_{dG} f^*(\Lambda_L[a]) \wedge d^c\psi \wedge \zeta$$

$$h(a,\zeta) = \int_{dG} f^*(\hat{\Lambda}_L[a]) \wedge d^c\psi \wedge \zeta$$

exist for all $a \in \mathfrak{A}^s_{p-1}(M,L,f)$. Hence for almost all $a \in G_{p-1}(V)$. Lemma 9.1 and (7.6) imply the existence of $I(h(\square,\zeta))$. Let $j: dG \longrightarrow \mathbb{R}(M)$ be the inclusion. If $\zeta > 0$, then $j^*(f^*(\hat{\Lambda}_L[a] \wedge d^c\psi \wedge \zeta) \geqq 0$. Assume f is holomorphic. Then (4.19), (4.17), Theorem 4.9 and Fubini's theorem imply (9.5). Especially, the right hand integral in (9.5) exists. Hence the right hand integral in (9.4) exist. Then (4.20), (4.17), Theorem 4.9 and Fubini's theorem imply (9.4). Now, (9.5) and (9.6) are extended to meromerphic f and complex valued ζ as in the proof of Lemma 9.1.

Now, (9.3) for $dd^c(\psi\zeta)$, (7.6) and (9.5) imply

$$\int_G f^*(c(L,\ell)^{p-1}) \wedge dd^c(\psi\zeta) = \frac{1}{p} I(h(\square,\zeta))$$

$$(9.7) \qquad = \int_{dG} f^*(c(L,\ell)^{p-1}) \wedge d^c\psi \wedge \zeta$$

which could also be proved by the use of Stokes Theorem. Now (9.7), and (7.5), (9.4) and (9.2) for $dd^c(\psi\zeta)$ imply (9.6); q.e.d.

Observe that Theorem 9.2 establishes also the existence of the integrals $\mu_p(L,f)$ and $\mu_p^0(L,f)$ in (8.14) and 8.15).

Lemma 9.3. (Crofton formula). Assume (A1) - (A6) and (A8) with $p \geqq 1$. Let ζ be a form of class C^2 and bidegree (q,q) on M. Let G be open on M such that $\overline{G} \cap \operatorname{supp} \zeta$ is compact. Let f be safe of order p for L. Then

$$(9.8) \qquad I\Big(\int_{G\cap f_a} \theta_f^a[L]\,\zeta\Big) = \int_G f^*(c(L,\ell)^p) \wedge \zeta.$$

Proof. Take a sequence $\{G_\nu\}_{\nu\in\mathbb{N}}$ of open subsets of G such that $\overline{G}_\nu$ is compact with $\overline{G}_\nu \subseteq G_{\nu+1} \subset G$, such that G_ν is a Stokes domain and such that $G = \bigcup_{\nu=1}^{\infty} G_\nu$. Take $\psi_\nu \colon M \longrightarrow \mathbb{R}[0,1]$ of class C^∞ such that $\psi_\nu \mid \overline{G}_\nu = 1$ and such that $\operatorname{supp} \psi_\nu \subset G_{\nu+1}$. Theorem 9.2 implies

$$I\Big(\int_{G\cap f_a} \psi_\nu\, \theta_f^a[L]\,\zeta\Big) = \int_G \psi_\nu\, f^*(c(L,\ell)^p) \wedge \zeta.$$

If $\zeta \geqq 0$, the Lebesgue bounded convergence Theorem implies (9.8). Observe that $f^*(c(L,\ell)^p) \wedge \zeta$ is integrable over G by Lemma 9.1. Hence (9.8) can be extended to complex valued ζ by the same method used in the proof of Lemma 9.1; q.e.d. (See [82], Lemma 10.4 and

Lemma 10.7 for more details)

Theorem 9.4. Assume (A1)-(A9) with $p \geqq 1$. Let f be safe of order p for L. Then

(9.9) $$I(m_a(L,f)) = c(p,k)\ \mu_p(L,f)$$

(9.10) $$I(m_a^0(L,f)) = c(p,k)\ \mu_p^0(L,f)$$

(9.11) $$I(D_a(L,f)) = c(p,k)\ \Delta_p(L,f)$$

(9.12) $$\Delta_p(L,f) = \mu_p(L,f) - \mu_p^0(L,f)$$

(9.13) $$I(N_a(L,f)) = T_p(L,f)$$

(9.14) $$I(n_a(L,f)) = A_p(L,f).$$

Proof. Lemma 9.3 implies (9.14) immediately. A function $\tilde{\psi}: M \longrightarrow \mathbb{R}$ of class C^2 exists such that $\tilde{\psi} \mid (\overline{G}-g) = \psi$. Then (9.6) for G with $\tilde{\psi}$ and g with $R - \tilde{\psi}$ holds. Subtraction gives (9.13). Also (9.2) for $K = G - \overline{g}$ and $\zeta = dd^c\tilde{\psi} \wedge \chi = dd^c\psi \wedge \chi$ on K gives (9.11). Let $j: dG \longrightarrow \mathfrak{R}(M)$ and $j_0: dg \longrightarrow \mathfrak{R}(M)$ be the inclusions. Then $j^*(d^c\tilde{\psi}) = j^*(d^c\psi)$ and $j_0^*(d^c(R - \tilde{\psi})) = -j_0^*(d^c\tilde{\psi})$. Hence (9.4) implies (9.9) and (9.10). Now, The First Main Theorem, (9.13), (9.9), (9.10) and (9.11) yield (9.12) q.e.d.

Lemma 9.5. Assume (A1)-(A6) and (A8) with $p \geqq 1$. Assume that f is safe of order p for L and that f is almost adapted of order p - 1 for L. Let $\zeta \geqq 0$ be a form of class C^2 and bidegree (q,q) on M. Assume $W \neq \emptyset$ is open in M with $\zeta > 0$ on W. Let I_f be the indeterminacy of f. Then the set W_0 of all $x \in \mathfrak{R}(W) - (f_\infty \cup I_f)$ such that $f^*(c(L,\ell)^p) \wedge \zeta$ is positive at x is open and dense in W.

Proof. Clearly, W_0 is open. It suffices to show that W_0 is dense in $W_1 = \mathfrak{R}(W) - (I_f \cup f_\infty)$. Let U be open in W_1. Let $G \neq \emptyset$ be open, such that $G = \mathrm{Int}\,\overline{G}$, such that $\overline{G}$ is compact and contained in U. By Proposition 5.5 and Theorem 5.7, $\mathfrak{A} = \mathring{\mathfrak{A}}^t_{p-1}(\overline{G},L,f)$ is open and non-empty. If $a \in \mathfrak{A}$, then $G \cap f_a \neq \emptyset$.

Then

$$h(a) = \int_{G \cap f_a} \theta_f^a[L]\ \zeta \geqq 0$$

exists for almost all $a \in G_{p-1}(V)$ and is positive on $\mathfrak{A}$. Hence

$$\int_G f^*(c(L,\ell)^p) \wedge \zeta = I(h) > 0.$$

Hence $W_0 \cap U \supseteq W_0 \cap G \neq \emptyset$; q.e.d.

Corollary 9.6. Assume (A1) - (A9) with $p \geqq 1$. Assume $\chi > 0$ (respectively $\psi\chi > 0$) on some non-empty, open subset of G. Assume that f is safe of order p for L and that f is almost adapted for L of order p - 1. Then $A_p(L,f) > 0$ (respectively $T_p(L,f) > 0$).

Compare the results of this section with Griffiths-King [32] 5(c).

IV. The First Main Theorem on parabolic spaces

10. Pseudo-convex and parabolic spaces

a) Exhaustion functions

Let M be a complex space of pure dimension $m > 0$. A proper map $\tau: M \longrightarrow \mathbb{R}_+$ of class C^∞ is called an <u>exhaustion</u> of M. If $A \subseteq M$ and $r \in \mathbb{R}_+$, define

$$A(r) = \{x \in A \mid \tau(x) < r^2\} \qquad A[x] = \{x \in A \mid \tau(x) \leqq r^2\}$$

$$A\langle r\rangle = \{x \in A \mid \tau(x) = r^2\} = A[r] - A(r).$$

$$A[s,r] = A[r] - A(s) \quad \text{if} \quad 0 \leqq s \leqq r.$$

Here, $M(r)$ is called the open <u>pseudo-ball</u>, $M[r]$ the <u>closed pseudo-ball</u> $M\langle r\rangle$ the <u>pseudo-sphere</u> all of radius r and $M[0]$ is called the <u>center</u>. Let $\mathfrak{G}_\tau$ be the set of all $r \in \overset{+}{\mathbb{R}}$ such that $M\langle r\rangle$ has locally finite $\mathfrak{H}^{2m-1}$-measure and such that $M\langle r\rangle - dM(r)$ has zero $\mathfrak{H}^{2m-1}$ measure. In this case write

$$\int_{M\langle r\rangle} = \int_{dM(r)}$$

Then $\overset{+}{\mathbb{R}} - \mathfrak{G}_\tau$ has measure zero. If τ is real analytic, then $\mathfrak{G}_\tau = \mathbb{R}^+$. Define $r_\tau^0 = \operatorname{Min}\{r \in \mathbb{R}_+ \mid M[r] \neq \emptyset\}$. The exhaustion τ is said to be <u>central</u> if $M[0]$ has measure zero, <u>strong</u> if $(dd^c\tau)^m \not\equiv 0$ on $M(r)$ for all $r > r_\tau^0$, <u>pseudo-convex</u> if $dd^c\tau \geqq 0$, <u>strictly pseudo-convex</u> if $dd^c\tau > 0$ on M, <u>logarithmic pseudo-convex</u> if $dd^c\log\tau \geqq 0$. A central, logarithmic pseudo-convex exhaustion τ is said to be <u>parabolic</u> if $(dd^c\tau)^m \not\equiv 0$ on

M and if

$$(dd^c \log \tau)^m \equiv 0 \qquad \text{on } M - M[0].$$

A central, logarithmic pseudo-convex exhaustion is pseudo-convex since $\tau\, dd^c \log \tau \leqq dd^c \tau$. A parabolic exhaustion is said to be strict if $dd^c \tau > 0$ on M. If τ is an exhaustion, define

(10.1) $$\upsilon = dd^c \tau \qquad \omega = dd^c \log \tau$$

(10.2) $$\rho_p = d^c \tau \wedge \upsilon^p \qquad \sigma_p = d^c \log \tau \wedge \omega^p$$

(10.3) $$\rho = \rho_{m-1} \qquad \sigma = \sigma_{m-1}$$

(10.4) $$\Phi(r) = \int_{M[r]} \upsilon^m \qquad \mathfrak{S}(r) = r^{-2m}\, \Phi(r)$$

Then

(10.5) $$d\rho = \upsilon^m \qquad d\sigma = \omega^m \qquad \tau^m \sigma = \rho$$

(10.6) $$\tau^{p+1} \omega^p = \tau\, \upsilon^p - p\, d\tau \wedge d^c \tau \wedge \upsilon^{p-1}$$

(10.7) $$\rho_p = d^c \tau \wedge \upsilon^p = \tau^{p+1} d^c \log \tau \wedge \omega^p = \tau^{p+1} \sigma_p$$

(10.8) $$d\tau \wedge \upsilon^p = \tau^{p+1}\, d \log \tau \wedge \omega^p.$$

If τ is pseudo-convex, then $\upsilon \geqq 0$ and $\Phi(r) \geqq 0$. If τ is also strong, then $\Phi(r) > 0$ for all $r > r_\tau^0$. If τ is logarithmic pseudo-convex then $\omega \geqq 0$. If τ is parabolic, then

(10.9) $$d\sigma = \omega^m = 0$$

(10.10) $$\tau\, \upsilon^m = m d\tau \wedge d^c \tau \wedge \upsilon^{m-1} = m d\tau \wedge \rho = m\, \tau^m d\tau \wedge \sigma.$$

If τ is a parabolic exhaustion on M, then (M, τ) is called a parabolic space. Now some properties of exhaustions shall be stated mostly without proof. These are left to the reader or he is referred to [82]§12. In this, M will denote a complex space of pure dimension m and τ an exhaustion on M. (For the concept of rank see [2]).

Lemma 10.1. Let $\varphi: N \longrightarrow M$ be a proper holomorphic map on a pure n-dimensional complex space N. Then $\tau\circ\varphi$ is an exhaustion. Moreover:

a) If τ is pseudo-convex, so is $\tau\circ\varphi$.

b) If τ is logarithmic pseudo-convex, so is $\tau\circ\varphi$.

c) If φ has strict rank m and if τ is central, then $\tau\circ\varphi$ is central.

d) If m = n, if φ has strict rank m and is surjective, and if τ is strong, then $\tau\circ\varphi$ is strong.

e) If m = n, if φ is locally biholomorphic and if φ is strictly pseudo-convex, then $\tau\circ\varphi$ is strictly pseudo-convex .

f) If m = n, if N is irreducible, if φ has strict rank m then $\Phi_{\tau\circ\varphi}(r) = s\Phi_\tau(r)$ where s is the sheet number of φ.(See [2] page 117).

g) If m = n, if φ has strict rank m and is surjective, if τ is parabolic, then $\tau\circ\varphi$ is parabolic.

Lemma 10.2. Define $\tau_0 = \sqrt{\tau}$. Let χ be a form of degree 2m - 1 on M(r) with r > 0 such that $d\tau_0 \wedge \chi$ is integrable over M(r), then χ is integrable over M <t> (i.e. dM(t)) for almost all $t \in \mathbb{R}(0,r)$ and

$$\int_{M(r)} d\tau_0 \wedge \chi = \int_0^r \Big(\int_{M\langle t\rangle} \chi \Big)\, dt .$$

Stokes Theorem and (10.5) and Lemma 7.2 imply easily.

Lemma 10.3. Take $r \in \mathfrak{G}_\tau$. Let j_r: $dM(r) \longrightarrow \mathfrak{R}(M)$ be the inclusion. Then

$$\Phi(r) = \int_{M\langle r\rangle} \rho \qquad\qquad \mathfrak{S}(r) = \int_{M\langle r\rangle} \sigma .$$

Moreover, if $x \in dM(r)$ and $\upsilon(x) \geqq 0$, then $j_r^*(\rho)(x) \geqq 0$ and $j_r^*(\sigma)(x) \geqq 0$. If $x \in dM(r)$ with $d\tau(x) \neq 0$ and $\upsilon(x) > 0$, then $j_r^*(\rho)(x) > 0$ and $j_r^*(\sigma)(x) > 0$.

<u>Lemma 10.4.</u> Let χ be locally integrable form of degree 2m on M. Define $S_r = \{x \in M\langle r\rangle \wedge \mathfrak{R}(M) \mid \chi(x) \neq 0\}$. Then $A(r) = \int_{M(r)} \chi$ is continuous from the left, and $B(r) = \int_{M[r]} \chi$ is continuous from the right. If S_r has measure zero at r, then $A(r) = B(r)$ at r and A and B are continuous at r.

The set $E = \{x \in \mathfrak{R}(M) \cap M\langle r\rangle \mid d((\tau - r^2)\chi)(x) \neq 0\}$ has measure zero by Tung [87] Theorem 7.1.1. Hence:

<u>Lemma 10.5.</u> Let χ be a form of degree $2m - 1$ and of class C^1 on M. Then a continuous function A on $\mathbb{R}^+$ is defined by

$$A(r) = \int_{M(r)} d\tau \wedge \chi = \int_{M[r]} d\tau \wedge \chi.$$

<u>Lemma 10.6</u>. Define $\tau_0 = \sqrt{\tau}$. Let $f\colon \mathbb{R}[s,r] \longrightarrow \mathbb{C}$ be of class C^1 with $0 < s \leqq r < \infty$. Let χ be a locally integrable form of degree 2m on M. Define $A(t) = \int_{M[t]} \chi$ for $t > 0$.

Then

$$\int_{M(s,r]} (f \circ \tau_0)\, \chi = f(r)\, A(r) - f(s)\, A(s) - \int_s^r A(t)\, f'(t)dt.$$

<u>Proposition 10.7.</u> Let τ be a logarithmic pseudo-convex exhaustion of M. Take $p \in \mathbb{Z}[0,m]$ and define $q = m - p$. Let $\chi \geqq 0$ be a form of bidegree (p,p) and class C^1 on M with $d\chi = 0$. For $r > 0$ define

$$A(r) = r^{-2q} \int_{M[r]} \chi \wedge \upsilon^q. \tag{10.11}$$

Then A increases and $A(0) = \lim_{0<r\to 0} A(r)$ exists. Moreover

(10.12) $$A(r) = \int_{M\langle r\rangle} \chi \wedge \sigma_{q-1} \qquad \text{if } q > 0 \text{ and } r \in \mathfrak{E}_\tau$$

(10.13) $$A(r) = \int_{M(0,r]} \chi \wedge \omega^q + A(0) \quad \text{if } r > 0.$$

If $q = 0$ and if τ is central, then $A(0) = 0$. If $p > 0$ and if $\upsilon > 0$ in a neighborhood of $M[0]$, then $A(0) = 0$ and $A(r) = O(r^{2p}\,\mathfrak{S}(r))$ for $r \longrightarrow 0$. If $p = 0$ and $\chi = 1$, then $A(r) = \mathfrak{S}(r)$ as defined in (10.4). Hence increases and $\mathfrak{S}(0) = \lim_{0<r\to 0} \mathfrak{S}(r)$ exists and

(10.14) $$\mathfrak{S}(r) = \int_{M(0,r]} \omega^m + \mathfrak{S}(0) \qquad \text{if } r > 0.$$

If τ is central, $M(0,r]$ can be replaced by $M[r]$ in (10.13) and in (10.14).

<u>Proof.</u> If $r \in \mathfrak{E}_\tau$, Stokes theorem implies

$$\int_{M\langle r\rangle} \chi \wedge \sigma_{q-1} = r^{-2q} \int_{M\langle r\rangle} \chi \wedge d^c\tau \wedge \upsilon^{q-1}$$

$$= r^{-2q} \int_{M(r)} \chi \wedge \upsilon^q = A(r).$$

If $0 < s < r \in \mathfrak{E}_\tau$ and $s \in \mathfrak{E}_\tau$, Stokes Theorem implies

(10.15) $$A(r) = \int_{M(s,r]} \chi \wedge \omega^q + A(s).$$

Semi-continuity from the right implies (10.15) for all $0 < s < r$. Hence A increases and $A(0)$ exists. Now $s \longrightarrow 0$ in (10.15) implies (10.13). If $q = 0$, then $p = m$. If τ is central, $M[0]$ has measure zero. Hence (10.11) implies $A(0) = 0$. If $p > 0$ and $\upsilon > 0$ on an open set $U \supset M[0]$, then $s > 0$ with $M[s] \subset U$ exists such that $0 \leqq \chi \leqq K\upsilon^p$ on $M(s)$ for some constant $K > 0$.

If $0 < r < s$, then

$$0 \leqq A(r) \leqq Kr^{-2q} \int_{M[r]} \upsilon^p \wedge \upsilon^q = Kr^{2p} \varsigma(r)$$

Now $\varsigma(r) \longrightarrow \varsigma(0)$ for $r \longrightarrow 0$. Hence $A(0) = 0$ since $p > 0$; q.e.d.

Theorem 10.8. Let τ be a central, logarithmic pseudo-convex exhaustion with $\omega^m \equiv 0$. Then the function ς defined in (10.4) is constant and

$$(10.16) \qquad \Phi(r) = \int_{M[r]} \upsilon^m = \int_{M(r)} \upsilon^m = \varsigma r^{2m}$$

Moreover, τ is parabolic if and only if $\varsigma > 0$. If τ is parabolic, then $M(r) \neq \phi$ for all $r > 0$ and $M[0] \neq \phi$.

Proof. Since $\omega^m \equiv 0$, the function ς is constant by (10.14). Also $\omega^m \equiv 0$ implies (10.10). Hence Lemma 10.5 implies (10.16). If τ is parabolic, then $M(r) \neq \phi$ and $\upsilon^m \not\equiv 0$ on $M(r)$ for some r. Hence $\varsigma > 0$. If $\varsigma > 0$, then $M(r) \not\equiv \phi$ and $\upsilon^m \not\equiv 0$ on $M(r)$ for each $r > 0$. Hence τ is parabolic. Assume that τ is parabolic and $M[0] = \phi$. Then $\tau > 0$ on M. Since τ is proper, a constant $s > 0$ exists such that $\tau > s^2$ on M. Hence $M(r) = \phi$ if $0 < r < s$, which is false. Hence $M[0] \neq \phi$. q.e.d.

If $c > 0$ and if τ is parabolic, then τ^c is parabolic. Hence τ can be normalized such that $\varsigma = 1$. Also Theorem 10.8 shows that a parabolic exhaustion is strong.

Lemma 10.9. If τ is a central, logarithmic pseudo-convex exhaustion with $\omega^m \equiv 0$, then $\varsigma(m!) = \int_M e^{-\tau}\upsilon^m$. Especially, this is true if τ is parabolic.

If τ is pseudo-convex, then $(dd^c\tau)^m > 0$ if and only if $dd^c\tau > 0$ at $x \in \Re(M)$. Hence

$$(10.17) \qquad \overset{+}{M} = \{x \in \Re(M) \mid \upsilon^m(x) > 0\} = \{x \in \Re(M) \mid \upsilon(x) > 0\}$$

is defined and open in M. If τ strong and pseudo-convex, then $\overset{+}{M}(r) \neq \phi$ for all $r > r_\tau^0$.

b) Examples of parabolic spaces

Parabolic exhaustions were introduced by Griffiths and King[33] who called them special exhaustions but who also suggested the name parabolic. However they require that the center consists of isolated points and that τ behaves like $|\mathfrak{g}|^2$ at these points where $\mathfrak{g}$ is a m-dimensional holomorphic vector function locally. Further they require that τ has only finitely many critical values. These requirements are unnecessary and excessively restrictive. A modification of a parabolic space may fail to be parabolic under these requirements. Also they needed deep results of Kodaira to construct special exhaustions on affine algebraic varieties. They require $\omega^{m-1} \not\equiv 0$ on the tangent space $T_z^{1,0}(dM(r))$ for some $z \in dM(r)$. This requirement is equivalent to $\upsilon^m \not\equiv 0$ on M as was shown in [82] Theorem 12.19. The detailed statement and the proof of the equivalence will be omitted here.

Obviously, $(\mathbb{C}^m, \tilde{\tau})$ with $\tilde{\tau}(\mathfrak{z}) = |\mathfrak{z}|^2$ is a parabolic space. Take $\mathfrak{a} \in \mathbb{C}^m$ and let $\beta: M \longrightarrow \mathbb{C}^m$ be a proper, holomorphic map of strict rank m on the pure m-dimensional complex space M. Define $\tau = |\beta - \mathfrak{a}|^2$. Then (M, τ) is a parabolic space and $\mathfrak{c}$ equals the sheet number. (Lemma 10.1g) and [82] Theorem 12.12). Such a parabolic space is said to be a cover of $\mathbb{C}^m$ and τ is called the covering exhaustion. Every affine algebraic variety of pure dimension m is a cover of $\mathbb{C}^m$. Hence an affine algebraic variety of pure dimension carries a parabolic exhaustion. (Griffiths-King [33] Proposition 2.4).

A Riemann surface M belongs to the class $\mathfrak{O}_G$ if and only if every subharmonic function which is bounded above is constant (Sario-Nakai [58] V7B). Assume M is not compact. Then M belongs to $\mathfrak{O}_G$ if and only if the following property holds:

(H) Let $G \neq \emptyset$ be open in M such that $\overline{G}$ is compact. Assume $dG = \partial G$. Assume $M - \overline{G}$ is connected. Let $\{G_\nu\}_{\nu \in \mathbb{N}}$ be a sequence of open subsets of M such that $\overline{G}_\nu$ is compact with $\partial G_\nu = dG_\nu$. Assume $\overline{G} \subset G_\nu \subset \overline{G}_\nu \subset G_{\nu+1}$. Assume $M = \bigcup_{\nu=1}^{\infty} G_\nu$. Let ω_ν be harmonic in $G_\nu - \overline{G}$ and continuous in $\overline{G}_\nu - G$ with $\omega_\nu \mid \partial G = 0$ and $\omega_\nu \mid \partial \overline{G}_\nu = 1$. Then $\omega_\nu(z) \longrightarrow 0$ for $\nu \longrightarrow \infty$ for all $z \in M - \overline{G}$. (Sario-Nakai [58] III 2F)

If (H) is satisfied for one such collection G, $\{G_\nu\}_{\nu\in\mathbb{N}}$, it is satisfied for all such collections.

Let M be a non-compact Riemann surface. Take $p \in M$. A harmonic function $h: M - \{p\} \longrightarrow \mathbb{R}$ is said to be on Evans-Selberg potential if for every $c \in \mathbb{R}$ a compact set K in M exists such that $h \geqq c$ on $M - K$ and if a smooth patch $\alpha: U_\alpha \longrightarrow U'_\alpha$ with $p \in U_\alpha$ and $\alpha(p) = 0$ and a harmonic function u on U_α exists such that $h = \log|\alpha| + u$ on $U_\alpha - \{p\}$. Then M belongs to $\mathfrak{O}_G$ if and only if there exists an Evans-Selberg potential on M.(Sario-Nakai [58] V13A]).

Define $M^2 = M \times M$ and $\Delta = \{(x,x) \mid x \in M\}$. A function $\eta: (M^2 - \Delta) \to \mathbb{R}$ is said to be an Evans-kernel if and only if the following conditions are satisfied:

(1) If $y \in M$, then η_y defined by $\eta_y(x) = \eta(x,y)$ on $M - \{y)$ is a Evans-Selberg potential.

(2) If $y_1 \in M$ and $y_2 \in M$, there exist a compact subset K of M containing y_1 and y_2 and a harmonic function h on $M - K$, bounded above, such that $\eta_{y_1} - \eta_{y_2} = h$ on $M - K$.

(3) If $(x,y) \in M^2 - \Delta$, then $\eta(x,y) = \eta(y,x)$.

Then M belongs to $\mathfrak{O}_G$ if and only if on Evans-kernel exists on M. (Sario-Nakai [58]V13E, actually they call $-\eta$ the Evans kernel). If η is an Evans kernel the following properties hold (Sario-Nakai [58] V13G):

(4) The function $\eta: (M^2 - \Delta) \longrightarrow \mathbb{R}$ is of class C^∞.

(5) If $z \in M$, then $\eta(x,y) \longrightarrow -\infty$ for $(x,y) \longrightarrow (z,z)$.

(6) If G is open and $\overline{G}$ is compact in M, if g is the Green function of G, then a function $v: G \times G \longrightarrow \mathbb{R}$ of class C^∞ exists such that $\eta = -g + v$ on $(G \times G) - \Delta$.

Actually, Sario-Nakai show only the continuity of η and v. However, η is separately harmonic, hence (6) follows applying the iterated Poisson formula locally in patches. Also $v = \eta + g$ is separately harmonic on $G^2 - \Delta$. Since v is continuous on G^2, v is separately harmonic and continuous on G^2, hence of class C^∞ on G^2.

A parabolic exhaustion τ of a complex space M is said to be simple if $M[0]$ consists of exactly one point and if $\varsigma = \varsigma_\tau = 1$. Again denote $M^2 = M \times M$ and $\Delta = \{(x,x) \mid x \in M\}$. A function $\tau: M^2 \longrightarrow \mathbb{R}$ of class C^∞

is called a parabolic kernel, if $\tau(x,y) = \tau(y,x)$ and if for each $y \in M$ the function τ_y defined by $\tau_y(x) = \tau(x,y)$ is a simple parabolic exaustion with center $\{y\}$.

Propositon 10.10. Let η be an Evans kernel on the non-compact Riemann surface M. Then $\tau = e^{2\eta}$ is a parabolic kernel on M where $\tau(x,x) = 0$ for all $x \in M$.

Proof. 1. Trivially τ is of class C^∞ on $M^2 - \Delta$. Take $z \in M$. A smooth patch $\alpha: U_\alpha \longrightarrow U'_\alpha$ exists with $z \in U_\alpha$ such that $\alpha(z) = 0 \in \mathbb{C}[1] \subset U'_\alpha$. Define $G = \alpha^{-1}(\mathbb{C}(1))$. By property (6) of η a function $v: G^2 \longrightarrow \mathbb{R}$ of class C^∞ exists such that

$$\tau(x,y) = \left|\frac{\alpha(x) - \alpha(y)}{1 - \alpha(x)\overline{\alpha(y)}}\right|^2 e^{2v(x,y)}$$

for all $(x,y) \in G \times G$. Hence $\tau: M^2 \longrightarrow \mathbb{R}_+$ is of class C^∞. Since $\eta(x,y) = \eta(y,x)$ if $x \neq y$, also $\tau(x,y) = \tau(y,x)$ for all $(x,y) \in M^2$ follows.

2. Take $y \in M$. Then η_y is an Evans-Selberg potential and τ_y is of class C^∞ on M. Take $r > 0$. A compact set K in M exists such that $\eta_y > \log r$ on $M - K$. Hence $\tau_y > r^2$ on $M - K$ and $M_{\tau_y}[r] \subseteq K$. Hence τ_y is an exhaustion with center $\{y\}$. Also η_y is harmonic on $M - \{y\}$. Hence $\omega = dd^c \log \tau_y = 2dd^c\eta_y = 0$ on $M - \{y\}$. Hence τ is a central logarithmic pseudo-convex exhaustion of M.

A smooth patch $\alpha: U_\alpha \longrightarrow U'_\alpha$ with $y \in U_\alpha$ and $\alpha(y) = 0$ and a harmonic function u on U_α exist such that $\eta_y = \log|\alpha| + u$ on U_α. Take $0 < r \in \mathfrak{C}_{\tau_y}$ such that $M_{\tau_y}[r] \subset U_\alpha$. Take $s > 0$ such that $\alpha(M_{\tau_y}(r)) \supset \mathbb{C}[s]$. Define $V = \alpha^{-1}(\mathbb{C}(s))$. Then

$$\varsigma = \int_{M\langle r\rangle} d^c \log \tau_y = \int_{dV} d^c \log \tau_y = 2\int_{dV} d^c\eta_y$$

$$= 2\int_{dV} d^c \log|\alpha| + 2\int_{dV} d^c u = 1$$

By Theorem 10.8, τ_y is parabolic and obviously simple. Hence τ is a parabolic kernel; q.e.d.

The same proof provides the following result:

Proposition 10.11. Let M be a non-compact Riemann surface. Take $y \in M$. Let $h: M - \{y\} \longrightarrow \mathbb{R}$ be an Evans-Selberg potential. Then $\tau = e^{2h}$ with $\tau(y) = y$ is a simple parabolic exhaustion of M with center $\{y\}$.

Theorem 10.12. A non-compact Riemann surface M has a parabolic exhaustion if and only if M belongs to $\mathfrak{D}_G$.

Remark. Classically the Riemann surfaces in $\mathfrak{D}_G$ are called parabolic. Hence Theorem 10.12 shows the consistency of the nomenclature in the non-compact case. However a compact Riemann surfaces is also parabolic in the classical sense but does not admit a parabolic exhaustion. For $dd^c\tau \geqq 0$ would imply that τ is constant which contradicts $dd^c\tau \not\equiv 0$.

Proof. If M belongs to $\mathfrak{D}_G$, an Evans-Selberg potential exists. By Proposition 10.11 a parabolic exhaustion exists. Assume a parabolic exhaustion τ is given on M. Take $r_n \in \mathbb{R}^+$ with $0 < r_0 < r_{n-1} < r_n \longrightarrow \infty$ for $r \longrightarrow \infty$ such that $dM(r_n) = M\langle r_n\rangle$. Define $s_n = \log r_n^2$ and

$$\omega_n = \frac{\log \tau - s_0}{s_n - s_0} \qquad \text{if } n \geqq 1.$$

Then ω_n is harmonic on $M(r_n) - M[r_0]$ with $\omega_n | M\langle r_0\rangle = 0$ and $\omega_n | M\langle r_n\rangle = 1$. If $z \in M - M[r_0]$, take $n_0 \in \mathbb{N}$ such that $x \in M(r_n)$ for all $n \geqq n_0$. Then $\log \tau(z) \geqq s_{n_0}$ and $s_n \longrightarrow \infty$ for $n \longrightarrow \infty$ imply $\omega_n(z) \longrightarrow 0$ for $n \longrightarrow \infty$. Condition (H) is satisfied and M belongs to $\mathfrak{D}_G$; q.e.d.

Theorem 10.13. Let (P, τ_1) and (Q, τ_2) be parabolic spaces of dimensions p and q respectively. Define $M = P \times Q$ and $m = p + q$. Let $\pi_1: M \longrightarrow P$ and $\pi_2: M \longrightarrow Q$ be the projections. Then $\tau = \tau_1 \circ \pi_1 + \tau_2 \circ \pi_2$

is a parabolic exhaustion with $\mathfrak{S}_\tau = \mathfrak{S}_{\tau_1} \cdot \mathfrak{S}_{\tau_2}$.

<u>Proof.</u> If $r \geqq 0$, then $M[r] \subseteq P[r] \times Q[r]$. Hence $M[r]$ is compact with $M[0] = P[0] \times Q[0]$. Hence τ is a central exhaustion. Abbreviate $u = \tau_1 \circ \pi_1$ and $w = \tau_2 \circ \pi_2$. Then $\tau = u + w$. If $\mu > p$ and $\nu > q$, then

$$(dd^c u)^\mu = 0 = (dd^c w)^\nu. \tag{10.18}$$

Because τ_1 and τ_2 are parabolic,

$$0 \leqq u^2 dd^c \log u = u\, dd^c u - du \wedge d^c u$$

$$0 \leqq w^2 dd^c \log w = w\, dd^c w - dw \wedge d^c w.$$

If f is any real valued function of class C^2, then $df \wedge d^c f \geqq 0$. Hence

$$0 \leqq u^2 w^2\, d \log \frac{u}{w} \wedge d^c \log \frac{u}{w}$$

$$= w^2 du \wedge d^c u + u^2 dw \wedge d^c w$$

$$- uw\, du \wedge d^c w - uw\, dw \wedge d^c u$$

$$\leqq w^2 u\, dd^c u + u^2 w\, dd^c w - uw(du \wedge d^c w + dw \wedge d^c u).$$

Therefore

$$uw\, \tau^2 dd^c \log \tau = uw(\tau\, dd^c \tau - d\tau \wedge d^c \tau)$$

$$= uw(u + w)\, dd^c(u + w) - uw\, d(u + w) \wedge d^c(u + w)$$

$$= uw(u\, dd^c u - du \wedge d^c u) + uw(w\, dd^c w - dw \wedge d^c w)$$

$$+ uw^2 dd^c u + u^2 w\, dd^c w - uw(du \wedge d^c w + dw \wedge d^c u) \geqq 0.$$

Hence $dd^c \log \tau \geqq 0$ almost every where on $M - M[0]$, hence everywhere on $M - M[0]$ by continuity. Therefore τ is logarithmic pseudo-convex.

Because τ_1 and τ_2 are parabolic,

$$u(dd^c u)^p = p\, du \wedge d^c u \wedge (dd^c u)^{p-1}$$

$$w(dd^c w)^q = q\, dw \wedge d^c w \wedge (dd^c w)^{q-1}$$

Also (10.18) implies

$$m(dd^c \tau)^{m-1} = m(dd^c u + dd^c w)^{m-1}$$

$$= \binom{m}{p}[p(dd^c u)^{p-1} \wedge (dd^c w)^q + q(dd^c u)^p \wedge (dd^c w)^{q-1}]$$

$$(dd^c \tau)^m = \binom{m}{p} (dd^c u)^p \wedge (dd^c w)^q.$$

Hence

$$m\, d\tau \wedge d^c \tau \wedge (dd^c \tau)^{m-1} = m\, d(u + w) \wedge d^c(u + w) \wedge (dd^c \tau)^{m-1}$$

$$= \binom{m}{p}[p\, du \wedge d^c u \wedge (dd^c u)^{p-1} \wedge (dd^c w)^q + q(dw \wedge d^c w) \wedge (dd^c u)^p \wedge dd^c w)^{q-1}]$$

$$= \binom{m}{p} (u + w)(dd^c u)^p \wedge (dd^c w)^q = \tau(dd^c \tau)^m.$$

Therefore

$$\tau^{m+1}(dd^c \log \tau)^m = \tau(dd^c \tau)^m - m\, d\tau \wedge d^c \tau \wedge (dd^c \tau)^{m-1} = 0.$$

Hence $(dd^c \log \tau)^m = 0$. Lemma 10.9 implies that ς is constant with

$$\sigma(m!) = \int_M e^{-\tau} \upsilon^m = \int_{P\times Q} e^{-u-w} \binom{m}{p} (dd^c u)^p \wedge (dd^c w)^q$$

$$= \binom{m}{p} \sigma_1(p!)\ \sigma_2(q!) = m!\ \sigma_1 \sigma_2.$$

Hence $\sigma = \sigma_1 \cdot \sigma_2 > 0$ and τ is parabolic;q.e.d.

Corollary 10.14. Let P and Q be complex spaces of pure dimensions $p > 0$ and $q > 0$ respectively. Let τ_1 and τ_2 be parabolic kernels on P and Q respectively. Define $M = P \times Q$ and $m = p + q$. Define $\tau: M^2 \longrightarrow \mathbb{R}_+$ by

$$\tau(x,y,w,z) = \tau_1(x,w) + \tau_2(y,z).$$

Then τ is a parabolic kernel on M.

By Theorem 10.13, $(\mathbb{C} - \mathbb{Z}) \times \mathbb{C}^m$ is parabolic. A.Andreotti pointed out to me that $(\mathbb{C} - \mathbb{Z}) \times \mathbb{C}^m$ is not affine algebraic.

Proposition 10.15. Let B be a compact complex space of pure dimension $m - 1$. Let $\pi: M \longrightarrow B$ be a non-positive holomorphic line bundle over B with a hermitian metric κ along the fibers of M such that $c(M,\kappa) \leqq 0$ on M. Assume an open subset $U \neq \emptyset$ of B exists such that $c(M,\kappa) < 0$ on U. Define $\ddot{\omega} = - c(M,\kappa) \geqq 0$. Define $\tau: M \longrightarrow \mathbb{R}_+$ by $\tau(z) = |z|_\kappa^2 = \kappa(z,z)$. Then τ is a parabolic exhaustion of M with

$$\sigma = \int_B \ddot{\omega}^{m-1} > 0.$$

Proof. Trivially τ is a central exhaustion with $dd^c \log \tau = \pi^*(\ddot{\omega}) \geqq 0$. Hence τ is logarithmic pseudo-convex. Also $(dd^c \log \tau)^m = \pi^*(\ddot{\omega}^m) = \pi^*(0) = 0$. Only $\sigma > 0$ remains to be shown. Define $\varphi = \pi | M(1)$. Pick $x_0 \in M\langle 1\rangle$ and define $j: \mathbb{C}(1) \longrightarrow \varphi^{-1}(\varphi(x_0))$ by $j(z) = zx_0$. Then j is biholomorphic. Let φ_* be the fiber integral to φ.

Then

$$\varphi_*(m\tau^{m-1}d\tau \wedge d^c\tau) = \int_{\mathbb{C}(1)} m|z|^{2m-2}d|z|^2 \wedge d^c|z|^2 = 1.$$

Therefore fiber integration implies.

$$\mathfrak{S} = \int_{M(1)} (dd^c\tau)^m = m \int_{M(1)} \tau^{m-1}d\tau \wedge d^c\tau \wedge \pi^*(\ddot{\omega}^{m-1})$$

$$= \int_B \varphi_*(m\tau^{m-1}d\tau \wedge d^c\tau)\ddot{\omega}^{m-1} = \int_B \ddot{\omega}^{m-1} > 0.$$

By Theorem 10.8 τ is a parabolic exhaustion; q.e.d.

Let V be complex vector space with a hermitian metric ℓ. Define $\tau\colon V \longrightarrow \mathbb{R}_+$ by $\tau(\mathfrak{z}) = |\mathfrak{z}|_\ell^2$. Let M be an analytic cone of pure dimension m in V with vertex 0; i.e. M is analytic in V with pure dimension m and $z \in \mathbb{C}$ and $\mathfrak{v} \in M$ implies $z\,\mathfrak{v} \in M$. Then $\tau|M$ is a parabolic exhaustion. This example is a special case of Proposition 10.15 up to a modification at 0. The easy proof is left to the reader. Observe $\upsilon > 0$ on M.

<u>Lemma 10.16.</u> Let τ be an exhaustion of the complex space M of pure dimension $m > 0$. Let $u\colon \mathbb{R}_+ \longrightarrow \mathbb{R}_+$ be a function of class C^∞ such that $u(x) \longrightarrow \infty$ for $x \longrightarrow \infty$. Then $u \circ \tau$ is an exhaustion with

$$\int_{M[r]} (dd^c(u\circ\tau))^m = (u'(r^2))^m \int_{M[r]} (dd^c\tau)^m.$$

Moreover, if $u' \geqq 0$ and $u'' \geqq 0$ and if τ is pseudo-convex, then $u\circ\tau$ is pseudo-convex.

For a proof see [80] Lemma 10.3 or [82] Lemma 12.20. When does u change τ into a parabolic exhausion $u\circ\tau$? A criterium is given in the following theorem whose proof shall be omitted (see [82] Theorem 12.21)).

Theorem 10.17. Let τ be a central exhaustion of the pure m-dimensional complex space M with $(dd^c\tau)^m \not\equiv 0$. Assume functions $g: \mathbb{R}_+ \longrightarrow \mathbb{R}_+$ and $u: \mathbb{R}_+ \longrightarrow \mathbb{R}_+$ of class C^∞ exist such that $g(x) > 0$ for all $x > 0$ and such that

$$d\tau \wedge d^c\tau \wedge (dd^c\tau)^{m-1} = (g\circ\tau)(dd^c\tau)^m$$

$$m(g\circ\tau)\, dd^c\tau \geqq d\tau \wedge d^c\tau$$

$$u(r) = \exp\left(\int_1^r \exp\left(-\int_1^t \frac{dx}{mg(x)}\right) dt\right) \qquad \text{if } r > 0.$$

Assume also $u(0) = 0$ and $u(r) \longrightarrow \infty$ for $r \longrightarrow \infty$. Then $u\circ\tau$ is a parabolic exhaustion.

Remark. g is unique if it exists and u is determined by g. However, $u(0)=0$, $u(r) \longrightarrow \infty$ for $r \longrightarrow \infty$ and u of class C^∞ are additional requirements on τ.

If τ is a parabolic exhaustion, then g and u with $g(x) = x/m$ and $u(x) = x$ satisfy the conditions of Theorem 10.17. Compare the results of this section with Griffiths-King [33] 2(a) and (b).

11. Jensen's formula on parabolic spaces

a) The counting function of analytic cycles

Let M be a complex space. Let $\mathfrak{A}_p$ be the set of all p-dimensional, irreducible analytic subsets of M. The support of a function $\nu: \mathfrak{A}_p \to \mathbb{Z}$ is the union of all $B \in \mathfrak{A}_p$ with $\nu(B) \neq 0$. Then ν is said to be an analytic cycle of dimension p if supp ν is empty or an analytic set of pure dimension p. The set $\mathfrak{Z}_p(M)$ of all analytic cycles of dimension p defines an ordered module over $\mathbb{Z}$. A function $\nu: M \to \mathbb{Z}$ is said to be an analytic precycle of dimension p if there exist analytic sets A and A^* such that $\dim A^* < p$, such that $A \subseteq \operatorname{supp} \nu \subseteq A \cup A^*$, such that $A = \emptyset$ or A is pure p-dimensional and such that ν is locally constant on $A - A^*$. Then $\operatorname{supp}_*\nu = A$ is unique and is called the proper support of ν. The set $\mathfrak{P}_p$ of all analytic precycle of dimension p on M is an ordered module and $\mathfrak{Q}_p = \{\nu \in \mathfrak{P}_p \mid \text{supp}_*\nu = 0\}$ is a sub-module. A homomorphism $j: \mathfrak{P}_p \to \mathfrak{Z}_p$ is defined. Take $\nu \in \mathfrak{P}_p$ with A and A* as above. Take $B \in \mathfrak{A}_p$. If $B \not\subseteq A$, define $j(\nu)(B) = 0$. If $B \subseteq A$, a constant $j(\nu)(B)$ exists such that $\nu(x) = j(\nu)(B)$ for all $x \in B - A^*$. Then j is surjective with $\ker j = \mathfrak{Q}_p$. Hence j factors to an isomorphism $\mathfrak{P}_p/\mathfrak{Q}_p \approx \mathfrak{Z}_p$ by which both modules are identified. Observe $\text{supp}_*\nu = \text{supp}\, j(\nu)$.

Let M be a complex space of pure dimension $m > 0$. Let τ be an exhaustion on M. Consider an analytic precycle ν of dimension p on M with $0 \leqq p \leqq m$. Define $A = \text{supp}_*\,\nu$. For $r > 0$ the counting function of ν is defined by

$$(11.1) \qquad n_\nu(r) = r^{-2p} \int_{A[r]} \nu\upsilon^p .$$

If $p = 0$, the integral is a sum. If A is a pure p-dimensional analytic subset on M or empty, let ν_A be the characteristic function of A. Then $\nu_A \in \mathfrak{P}_p$ and $n_A = n_{\nu_A}$ is defined. If $\nu \in \mathfrak{P}_p$, then

$$(11.2) \qquad n_\nu(r) = \sum_{B\in\mathfrak{A}_p} j(\nu)(B)\, n_B(r).$$

Then n_ν is additive in ν on $\mathfrak{P}_p$ and factors to a homomorphism $\mathfrak{Z}_p \longrightarrow \mathbb{R}$. If $\nu \in \mathfrak{P}_p$ and $A = \text{supp}_*\,\nu$, then $\tau \mid A$ is an exhaustion on A. The result of §10 become applicable. Hence n_ν is continuous from the right and continuous at every $r \in \mathfrak{C}_{\tau/A}$ with

$$(11.3) \qquad n_\nu(r) = \int_{A\langle r\rangle} \nu\, \sigma_{p-1} = r^{-2p} \int_{A\langle r\rangle} \nu\, \rho_{p-1}.$$

If $0 < s < r$ and if $f: \mathbb{R}[s,r] \longrightarrow \mathbb{C}$ is of class C^1, then

(11.4) $$\int_{A(s,r]} \nu\, f(\sqrt{\tau})\upsilon^p = r^{2p} n_\nu(r) f(r) - s^{2p} n_\nu(s) f(s) - \int_s^r t^{2p} n_\nu(t) f'(t)\,dt$$

(11.5) $$\int_{A(s,r]} \nu f(\sqrt{\tau})\omega^p = n_\nu(r) f(r) - n_\nu(s) f(s) - \int_s^r n_\nu(t) f'(t)\,dt.$$

If τ is logarithmic pseudo-convex, then $n_\nu(r) \longrightarrow n_\nu(0)$ for $r \longrightarrow 0$ and

(11.6) $$n_\nu(r) = \int_{A(0,r]} \nu\, \omega^p + n_\nu(0).$$

If $\nu \geqq 0$ and if τ is pseudo-convex, then $n_\nu \geqq 0$. If $\nu \geqq 0$ and if τ is logarithmic pseudo-convex, then $n_\nu \geqq 0$ increases.

For $r > 0$ and $p \in \mathbb{N}$, define φ_r and ψ_r^p by $\varphi_r(x) = \log^+ r/x$ and $\psi_r^p(x) = \frac{1}{2p}(x^{-2p} - r^{-2p})$ for $x > 0$. Then

(11.7) $$2\, d^c(\psi_r^p \circ \sqrt{\tau}) \wedge \upsilon^p = -\, d^c \log \tau \wedge \omega^p = -\,\sigma_p$$

(11.8) $$2\, dd^c(\psi_r^p \circ \sqrt{\tau}) \wedge \upsilon^p = -\,\omega^{p+1}$$

on $M - M[0]$. If $0 < s < r$, define φ_{sr} and ψ_{sr}^p by $\varphi_{sr}(x) = \mathrm{Min}\,(\varphi_r(s), \varphi_r(x))$ and $\psi_{sr}^p(x) = \mathrm{Min}\,(\psi_r^p(s), \psi_s^p(x))$. Then φ_{sr} and ψ_{sr}^p are constant on $\mathbb{R}[0,s]$ and $\varphi_{sr} = \varphi_r$ and $\psi_{sr}^p = \psi_r^p$ on $\mathbb{R}[s,\infty)$.

The <u>valence function</u> of a precycle ν of dimension p is defined for $0 < s < r$ by

(11.9) $$N_\nu(r,s) = \int_s^r n_\nu(t)\,\frac{dt}{t}.$$

Define $A = \text{supp}_* \nu$. If $A[0] = \phi$, then $n_\nu = 0$ for sufficiently small $r > 0$ and $N_\nu(r) = N_\nu(r,0)$ is defined by (11.9) for $s = 0$. The function N_ν is continuous. If τ is logarithmic pseudo-convex and if $0 < s < r$, then (11.4) and (11.5) show easily

$$(11.10) \qquad N_\nu(r,s) = \int_{A(0,r]} \nu(\varphi_{sr} \circ \sqrt{\tau})\, \omega^p + \varphi_{sr}(0)\, n_\nu(0).$$

$$(11.11) \qquad N_\nu(r,s) = \int_{A[r]} \nu(\psi^p_{sr} \circ \sqrt{\tau})\, \upsilon^p \qquad \text{if } p > 0.$$

The origins of (11.10) and(11.11) go back to H. Kneser [39].

b) Green's Residue formula and Jensen's formula

Theorem 11.1. (Green's Residue formula) Let τ be an exhaustion of the complex space M of pure dimension $m > 0$. Let $f\colon M \longrightarrow \mathbb{C}^n$ a p-fibering holomorphic map with $p = m - n$. Abbreviate $\nu = \nu^0_f$ and $A = f^{-1}(0)$. Define $\hat{\tau}\colon \mathbb{C}^n \longrightarrow \mathbb{R}_+$ by $\hat{\tau}(\mathfrak{z}) = |\mathfrak{z}|^2$ and $\hat{\omega} = dd^c \log \hat{\tau}$. Take $s \in \mathfrak{C}_\tau$ and $r \in \mathfrak{C}_\tau$ with $0 < s < r$. Then

(11.12)

$$N_\nu(s,r) = \int_{M\langle r\rangle} \log |f|\, f^*(\hat{\omega}^{n-1}) \wedge \sigma_p - \int_{M\langle s\rangle} \log |f|\, f^*(\hat{\omega}^{n-1}) \wedge \sigma_p$$

$$+ \int_{M(s,r]} \log |f|\, f^*(\hat{\omega}^{n-1}) \wedge \omega^{p+1}.$$

(Compare Griffiths-King [33] Proposition 3.17).

Proof. Here ν is an analytic precycle of dimension p with $A = \text{supp}_* \nu = \text{supp}\, \nu$. Take $\lambda\colon M \longrightarrow \mathbb{R}[0,1]$ of class C^∞ with $\lambda(x) = 0$ if $x \in M[s/4]$ and $\lambda(x) = 1$ if $x \in M - M[s/2]$. Assume $p > 0$. If $t > 0$, define $\xi_t = \lambda\, \psi^p_t \circ \sqrt{\tau}$ on $M - M[0]$ and $\xi_t = 0$ in $M[0]$.

Then ξ_t is of class C^∞ with $\xi_r = \psi^p_{sr}\circ\sqrt{\tau}$ on $M(s,r]$. Also $2\, d^c\xi_t \wedge \upsilon^p = -\sigma_p$ on $M(s/2,r]$. Define $\zeta = (1-\lambda)\, \psi^p_r(s)$. Then

$\xi_r - \xi_s + \zeta = \psi^p_{sr}\circ\sqrt{\tau}$ on $M[s]$. Hence $2\, dd^c\xi_r \wedge \upsilon^p = -\omega^{p+1}$ on $M(s,r]$ and $dd^c(\xi_r - \xi_s + \zeta) = 0$ on $M(s)$. Theorem A8 (Appendix) implies

$$\int_{M(r)} \log |f|^2 f^*(\hat{\omega}^{n-1}) \wedge dd^c\xi_r \wedge \upsilon^p$$

$$= \int_{A(r)} \nu\, \xi_r \upsilon^p - \int_{M\langle r\rangle} \log |f|^2 f^*(\hat{\omega}^{n-1}) \wedge \sigma_p$$

$$\int_{M(s)} \log |f|^2 f^*(\omega^{n-1}) \wedge dd^c\xi_s \wedge \upsilon^p$$

$$= \int_{A(s)} \nu\, \xi_s\, \upsilon^p - \int_{M\langle s\rangle} \log |f|^2 f^*(\hat{\omega}^{n-1}) \wedge \sigma_p$$

$$\int_{M(s)} \log |f|^2 f^*(\hat{\omega}^{n-1}) \wedge dd^c\zeta \wedge \upsilon^p = \int_{A(s)} \nu\, \zeta\, \upsilon^p.$$

Addition implies (11.12) for $p > 0$. If $p = 0$, define $\xi_t = \lambda\, \varphi_t\circ\sqrt{\tau}$ and $\zeta = (1 - \lambda)\, \varphi_r(s)$ and proceed as above; q.e.d.

Using Theorem 7.4, the same method gives the following result:

Theorem 11.2 (Jensen Formula). Let τ be an exhausion of the complex space M of pure dimension m. Let f be a meromorphic function on M such that $f^{-1}(0)$ is thin, Let θ_f be the divisor multiplity of f. Take $r \in \mathfrak{E}_\tau$ and $s \in \mathfrak{E}_\tau$ with $0 < s < r$. Then

(11.13)

$$N_{\theta_f}(r,s) = \int_{M\langle r\rangle} \log |f|\sigma - \int_{M\langle s\rangle} \log |f|\sigma - \int_{M(s,r]} \log |f|\omega^m.$$

If τ is parabolic, then

$$(11.14)\qquad N_{\theta_f}(r,s) = \int_{M\langle r\rangle} \log|f|\,\sigma - \int_{M\langle s\rangle} \log|f|\,\sigma$$

(Compare Griffiths-King [33] Proposition 3.2)

Theorem 11.3. Let (M,τ) be an irreducible parabolic space. Let f be a meromorphic function on M. Then:

Liouville: If f is bounded, f is constant.

Casorati-Weierstrass: If f is not constant, f(M) is dense in $\mathbb{P}_1$.

Small Picard: If M is normal and simply connected and if f omits at least three values in $\mathbb{P}_1$, then f is constant.

Proof. At first Liouville's theorem shall be proved. Let f be a bounded meromorphic function. Assume f is not constant. Define M^+ by (10.17). Then $x \in M^+$ exists since τ is parabolic. Define $a = f(x)$. Then $h = f - a$ is bounded, non-constant and meromorphic. Hence $A = h^{-1}(0)$ is thin and $\nu = \theta_h \geqq 0$ exists. Take $s > 0$ such that $x \in M(s)$ Then $x \in A \cap M^+ \cap M(s)$. Hence $n_\nu(s) > 0$. A constant $c > 0$ exists such that $|h| \leqq c$ on M. Define

$$c_1 = \varsigma \log c - \int_{M\langle s\rangle} \log|h|\,\sigma.$$

Take $r \in \mathfrak{C}_\tau$ such that $n_\nu(s)\log\frac{r}{s} > c_1$. Now (11.14) implies $N_{\theta_h}(r,s) \leqq c_1$. Hence

$$c_1 < n_\nu(s)\log\frac{r}{s} \leqq \int_s^r n_\nu(t)\,\frac{dt}{t} \leqq c_1.$$

Contradiction! Liouville's theorem is proved and implies the Casorati-Weierstrass Theorem trivially. If M is normal a point x of indetermina[cy] blows up to $f(x) = \mathbb{P}_1$. Hence, if f omits values there are no indetermi[n]acies. Now the small Picard theorem is proved along standard lines; q.e[d].

(Compare Griffiths-King [33]. Proposition 3.8.)

Lemma 11.4. Take $s > 0$. Assume $g: \mathbb{R}_+ \longrightarrow \mathbb{R}_+$ increases. Define $0 \leqq g(\infty) = \lim_{r\to\infty} g(r) \leqq \infty$. For $r \geqq s$ define

$$h(r) = \int_s^r g(t) \frac{dt}{t}$$

Then $h(r)/\log r \longrightarrow g(\infty)$ for $r \longrightarrow \infty$.

Let τ be a logarithmic pseudo-convex exhaustion of M. Let $\nu \geqq 0$ be an analytic precycle of dimension p on M with $0 \leqq p \leqq m$. Then $n_\nu \geqq 0$ increases and $n_\nu(\infty) = \lim_{r\to\infty} n_\nu(r)$ is defined. Then ν is said to have rational growth if $n_\nu(\infty) < \infty$ and to have transcendental growth if $n_\nu(\infty) = \infty$. Lemma 11.4 implies

$$\lim_{r\to\infty} \frac{N_\nu(r,s)}{\log r} = n_\nu(\infty). \tag{11.15}$$

Hence ν has rational growth if and only if $N_\nu = O(\log r)$ for $r \longrightarrow \infty$.

Compare the results of this section with Griffiths-King [33] 3(a) and (c).

12. The First Main Theorem on logarithmic pseudo-convex spaces

a) Discussion of the value distribution functions

The following general assumptions shall be made:

(B1) Let τ be a logarithmic pseudo-convex exhaustion of the complex space M of pure dimension $m > 0$.

(B2) Let L be a holomorphic line bundle over the complex space N. Let κ be a hermitian metric along the fibers of L.

(B3) Let V be a complex vector space of dimension $k + 1 \geqq 1$ with a hermitian metric ℓ.

(B4) Let $\eta: N \times V \longrightarrow L$ be a semi-amplification of L.

(B5) Let $f: M \longrightarrow N$ be a meromorphic map.

(B6) Take $p \in \mathbb{Z}[0,k]$ with $q = m - p \geqq 0$.

(B7) Assume that $c(L,\kappa) \geqq 0$.

(B8) Assume that κ is distinguished.

(B9) Assume that f is safe for L.

Remark 12.1. If (B1) - (B6) hold and if $s \in \mathfrak{C}_\tau$ and $r \in \mathfrak{C}_\tau$ with $0 < s < r$, then (A1) - (A9) are satisfied with $g = M(s)$ and $G = M(r)$ and $\psi = \psi^q_{sr} \circ \sqrt{\tau}$ and $\chi = \upsilon^q$.
Moreover, on M[s,r] we have

$$(12.1) \qquad 2\, d^c\psi \wedge \chi = -\sigma_q \qquad\qquad 2\, dd^c\psi \wedge \chi = -\omega^{q+1}.$$

Assume (B1), (B2) and (B5). Take $p \in \mathbb{Z}[0,m]$ and define $q = m - p \geqq 0$. By Lemma 7.3 the spherical image of order p of f for (L,κ)

$$(12.2) \qquad A_p(r) = A_p(r,L,\kappa,f) = r^{-2q} \int_{M[r]} f^*(c(L,\kappa)^p) \wedge \upsilon^q$$

exists for $r > 0$ and is semi-continuous from the right. If $r \in \mathfrak{C}_\tau$, then

$$(12.3) \qquad r^{2q} A_p(r) = A_p(L,\kappa,f)$$

under the Remark 12.1 translation. Observe

$$(12.4) \qquad A_0(r) = \mathfrak{S}(r) = r^{-2m}\Phi(r).$$

Lemma 12.2. Assume (B1),(B2),(B5) and (B7). Take $p \in \mathbb{Z}[1,m]$ and define $q = m - p$. Then $A_p \geqq 0$ increases and $A_p(0) = \lim_{0<r\to 0} A_p(r)$ exists with

(12.5)
$$A_p(r) = \int_{M(0,r]} f^*(c(L,\kappa)^p) \wedge \omega^q + A_p(0)$$

(12.6)
$$A_p(r) = \int_{M\langle r\rangle} f^*(c(L,\kappa)^p) \wedge \sigma_{q-1} \quad \text{if } r \in \mathfrak{C}_\tau.$$

If τ is central, then $A_m(0) = 0$. If $r_1 > 0$ exists such that either $M(r_1) = \phi$ or such that $\upsilon > 0$ on $M(r_1)$ and that f is holomorphic on $M(r_1)$, then $A_p(r) = O(r^{2p} \cdot \mathfrak{S}(r))$ for $r \longrightarrow 0$. Hence $A_p(0) = 0$.

Proof. $c(L,\kappa) \geqq 0$ is assumed. If f is holomorphic, Proposition 16.7 implies Lemma 12.2. If f is meromorphic, go over the graph; q.e.d.

Assume (B1) - (B5) and (B9). Take $p \in \mathbb{Z}[0,m]$. Define $q = m - p$. For $r > 0$ the spherical image of order p of f for L is defined by

(12.7)
$$A_p[r] = A_p[r,L,f] = r^{-2q} \int_{M[r]} f^*(c(L,\ell)^p) \wedge \upsilon^q \geqq 0.$$

If $r \in \mathfrak{C}_\tau$, then $r^{2q}A_p[r] = A_p(L,f)$ by Remark 12.1. Hence $A_p[r]$ exists for all $r \in \mathfrak{C}_\tau$, hence for all $r > 0$ and increases and is semi-continuous from the right. Hence $A_p[0] = \lim_{0<r\to 0} A_p[r]$ exists. If $p = 0$, then $A_0[r] = \mathfrak{S}(r)$. If $p > 0$ and $r > 0$, then

(12.8)
$$A_p[r] = \int_{M(0,r]} f^*(c(L,\ell)^p) \wedge \omega^q + A_p[0]$$

(12.9)
$$A_p[r] = \int_{M\langle r\rangle} f^*(c(L,\ell)^p) \wedge \sigma_{q-1} \quad \text{if } r \in \mathfrak{C}_\tau$$

and the rest of Lemma 12.2 also holds. This is easily seen by considering the meromorphic map $\varphi \circ f$ and the hyper-plane section bundle instead of L. (See [82] Lemma 14.3).

Assume (B1), (B2) and (B5). Take $p \in \mathbb{Z}[0,m]$. Define $q = m - p$. The characteristic function of order p of f for (L,κ) is defined for $0 < s < r$ by

$$(12.10) \qquad T_p(r,s) = T_p(r,s,L,\kappa,f) = \int_s^r A_p(t) \frac{dt}{t}.$$

If the integral exists for $s = 0$, define also $T_p(r) = T_p(r,0)$ by (12.10) for $s = 0$. If $p = 0$, then $T_0(r,s) = \int_s^r \varsigma(t)\, dt/t$. If $p \geqq 1$, Lemma 10.6 implies

$$(12.11) \qquad T_p(r,s) = \int_{M(0,r]} (\psi_{sr}^q \circ \sqrt{\tau})\, f^*(c(L,\kappa)^p) \wedge \upsilon^q.$$

Hence if $r \in \mathfrak{G}_\tau$ and $s \in \mathfrak{G}_\tau$, then $T_p(r,s) = T_p(L,\kappa,f)$ in the translation of Remark 12.1. If also (B7) holds, then Lemma 10.6 and (12.5) imply

$$(12.12) \qquad T_p(r,s) = \int_{M(0,r]} (\varphi_{sr} \circ \sqrt{\tau})\, f^*(c(L,\kappa)^p) \wedge \omega^q + \varphi_{sr}(0)\, A_p(0)$$

and $T_p(r,s) \geqq 0$ increases.

Assume (B1) - (B5) and (B9). Take $p \in \mathbb{Z}[0,m]$. Define $q = m - p$. For $0 < s < r$, the characteristic function of order p of f for L is defined by

$$(12.13) \qquad T_p[r,s] = T_p[r,s,L,f] = \int_s^r A_p[t] \frac{dt}{t} \geqq 0.$$

Define $T_p[r] = T_p[r,0]$ by (12.13) for $s = 0$ if the integral exists.

If $p = 0$, then $T_0[r,s] = \int_s^r \mathfrak{S}(t)\, dt/t$. If $p \geqq 1$, then

(12.14) $$T_p[r,s] = \int_{M(0,r]} (\psi_{sr} \circ \sqrt{\tau})\, f^*(c(L,\ell)^p) \wedge \upsilon^q.$$

(12.15)
$$T_p[r,s] = \int_{M(0,r]} (\varphi_{sr} \circ \sqrt{\tau})\, f^*(c(L,\ell)^p) \wedge \omega^q + \varphi_{sr}(0)\, A_p(0).$$

If $r \in \mathfrak{E}_\tau$ and $s \in \mathfrak{E}_\tau$, then $T_p[r,s] = T_p(L,f)$ in the translation of Remark 12.1.(For more details see [82]§14). If $p = 1$, write also

(12.16) $$A(r) = A_1[r] \qquad A[r] = A_1[r]$$

(12.17) $$T(r,s) = T_1(r,s) \qquad T[r,s] = T_1[r,s].$$

Assume (B1) - (B6) with $p \geqq 1$. Take $a \in G_{p-1}(V)$. Assume that f is adapted to a for L. Then $\nu = \theta_f^a[L]$ is an analytic precycle of dimension q on M. Define the counting function of f to a for L by $n_a(r) = n_a(r,L,f) = n_\nu(r)$ and the valence function by $N_a(r,s) = N_a(r,s,L,f) = N_\nu(r,s)$. For the properties of these functions see section 11. If $r \in \mathfrak{E}_\tau$ and $s \in \mathfrak{E}_\tau$, then $n_a(r) = n_a(L,f)$ and $N_a(r,s) = N_a(L,f)$ in the sense of Remark 12.1 and (8.6) and (8.7).

Assume (B1) - (B6) with $p \geqq 1$. Take $a \in G_{p-1}(V)$. Assume that f is adapted to a for L. For $0 < s < r$, the deficit of f to a for (L,κ) is defined by

(12.18) $$D_a(r,s) = D_a(r,s,L,\kappa,f) = \tfrac{1}{2} \int_{M(s,r]} f^*(\Lambda_L[a]_\kappa) \wedge \omega^{q+1}.$$

Then (11.8) implies

(12.19) $$D_a(r,s) = \int_{M(s,r]} f^*(\Lambda_L[a]_\kappa) \wedge d^c d(\psi_{sr}^q \circ \sqrt{\tau}) \wedge \upsilon^q.$$

Hence $D_a(r,s) = D_a(L,\kappa,f)$ if $r \in \mathfrak{G}_\tau$ and $s \in \mathfrak{G}_\tau$ by Remark 12.1 and (8.10). Hence the integral exists in this case and consequently for all $0 < s < r$. If also (B7) and (B8) hold, then $\Lambda_L[0]_\kappa \geqq 0$. Hence $D_a(r,s) \geqq 0$ increases in r. If τ is parabolic and if $a \in \mathbb{P}(V)$ i.e. if $p = 1$ then $\omega^m \equiv 0$ and

$$(12.20) \qquad D_a(r,s) = 0.$$

The importance of the parabolic exhaustion consists in the vanishing of the deficit.

Assume (B1) - (B6) with $p \geqq 1$. Take $a \in G_{p-1}(V)$. Assume that f is safe of order p and that f is strictly adapted to a. For $0 < s < r$, the deficit of f to a for L is defined by.

$$(12.21) \qquad D_a[r,s] = D_a[r,s,L,\ell,f] = \frac{1}{2}\int_{M(s,r]} f^*(\Lambda_L[a]) \wedge \omega^{q+1}.$$

$$(12.22) \qquad D_a[r,s] = \int_{M(s,r]} f^*(\Lambda_L[a]) \wedge d^c d(\psi^q_{s,r} \circ \sqrt{\tau}) \wedge \upsilon^q.$$

Here $D_a[r,s] \geqq 0$ increases in r. If $r \in \mathfrak{G}_\tau$ and $s \in \mathfrak{G}_\tau$ then $D_a[r,s] = D_a(L,f)$, which implies, that $D_a[r,s]$ exists for all $0 < s < r$. If $p = 1$, and if τ is parabolic, then

$$(12.23) \qquad D_a[r,s] = 0.$$

Assume (B1) - (B6) with $p \geqq 1$. Take $a \in G_{p-1}(V)$. Assume that f is adapted to a for L. For $r \in \mathfrak{G}_\tau$ the compensation function of f for (L,κ) to a is defined by

$$(12.24) \qquad m_a(r) = m_a(r,L,\kappa,f) = \frac{1}{2}\int_{M\langle r\rangle} f^*(\Lambda_L[a]_\kappa) \wedge \sigma_q.$$

Then $m_a(r) = m_a(L,\kappa,f)$. Hence the integral exists. Also $m_a(s) = m_a^0(L,\kappa,f)$. If (B7) and (B8) hold, then $\Lambda_L[a]_\kappa \geqq 0$, hence $m_a(r) \geqq 0$.

Assume that f is safe of order p and that f is strictly adapted to a. For $r \in \mathfrak{G}_\tau$, the <u>compensation function</u> of f for L to a is defined by

$$(12.25) \qquad m_a[r] = m_a[r,L,f] = \frac{1}{2}\int_{M\langle r\rangle} f^*(\Lambda_L[a]) \wedge \sigma_q .$$

Then $m_a[r] = m_a(L,f)$ and $m_a[s] = m_a^0(L,f)$ if $s \in \mathfrak{G}_\tau$. So the integral $m_a[r]$ exists. Observe $m_a[r] \geqq 0$. If $p = 1$ and $r \in \mathfrak{G}_\tau$ then

$$(12.26) \qquad m_a(r) = \int_{M\langle r\rangle} \log \frac{1}{||f,a||_\kappa} \sigma$$

$$(12.27) \qquad m_a[r] = \int_{M\langle r\rangle} \log \frac{1}{||f,a||_\ell} \sigma \geqq 0.$$

If κ is distinguished, then $m_a(r) \geqq 0$ even if (B7) fails.

b) <u>The First Main Theorem</u>

Remark 12.1 and Theorem 8.1 imply immediately:

<u>Theorem 12.3.</u> (<u>First Main Theorem</u>). Assume (B1) - (B6) with $p \geqq 1$. Take $\mathbf{a} \in G_{p-1}(V)$. Let f be adapted to a for L.Take $r \in \mathfrak{G}_\tau$ and $s \in \mathfrak{G}_\tau$ with $0 < s < r$. Then

$$(12.28) \qquad T_p(r,s) = N_a(r,s) + m_a(r) - m_a(s) - D_a(r,s).$$

If $p = 1$ and if τ is parabolic, then

$$(12.29) \qquad T(r,s) = N_a(r,s) + m_a(r) - m_a(s).$$

If f is safe of order p and if f is strictly adapted to a for L, then

$$(12.30) \qquad T_p[r,s] = N_a(r,s) + M_a[r] - m_a[s] - D_a[r,s].$$

If in addition $p = 1$ and if τ is parabolic, then

$$(12.31) \qquad T[r,s] = N_a(r,s) + m_a[r] - m_a[s].$$

Keep $s \in \mathfrak{C}_\tau$. Then T_p, N_a and D_a are semi-continuous from the right for all $r > s$. Since s arbitrary, m_a extends to a function semi-continuous from the right and defined for all $r > 0$ such that the First Main Theorem holds. Now s can be freed, and right semi-continuity extends the First Main Theorem to all $r > s > 0$.

The First Main Theorem can also be extended to the case $s = 0$; see [82]§14.

c) Integral averages

Again I denotes the integral average operator on $G_{p-1}(V)$ and $c(p,k)$ is defined by 9.1.

Theorem 12.4. Assume (B1) - (B6). Assume that f is safe of order p for L. Take $0 < s < r$. Then $m_a[r]$, $D_a[r,s]$, $n_a[r]$ and $N_a[r,s]$ are defined for almost all $a \in G_{p-1}(V)$ and

$$(12.32) \qquad A_p[r] = I(n_a(r))$$

$$(12.33) \qquad T_p[r,s] = I(N_a(r,s))$$

$$(12.34) \qquad I(m_a[r]) = \tfrac{1}{2}c(p,k)\, A_{p-1}[r]$$

$$(12.35) \qquad I(D_a[r,s]) = \tfrac{1}{2}\, c(p,k)(A_{p-1}[r] - A_{p-1}[s]).$$

Especially, if p = 1, then

$$I(m_a[r]) = \tfrac{1}{2}\Big(\sum_{\mu=1}^{k} \frac{1}{\mu}\Big)\big(\mathfrak{S}(r) - \mathfrak{S}(s)\big) \tag{12.35}$$

$$I(D_a[r,s]) = \tfrac{1}{2}\Big(\sum_{\mu=1}^{k} \frac{1}{\mu}\Big)\, \mathfrak{S}(r). \tag{12.37}$$

Proof. By right semi-continuity only $r \in \mathfrak{E}_\tau$ and $s \in \mathfrak{E}_\tau$ have to be considered. Remark 12.1 and Theorem 9.4 imply (12.32) and (12.33) immediately. Also (8.5), (12.1) and (12.8) show

$$2\Delta_p(L,f) = A_{p-1}[r] - A_{p-1}[s].$$

Hence (9.11) implies (12.35). Similar, (8.14), (12.1) and (12.6) show $2\mu_p(L,f) = A_p[r]$. Hence (9.9) implies (12.34), q.e.d.

The averages (12.34) and (12.35) are surprising but are no accident. In a much deeper sense, the same principle holds for Schubert cycles on Grassmann manifolds and Schubert zeros in vector bundles. See Matsushima [47] and Stoll [84].

d) Dependence on the hermitian metric κ

The dependency of the value distribution function on (L,κ) is studied in Griffiths-King [33] 5 and in [82]§14d. Here only the change of κ shall be considered.

Theorem 12.5. Assume (B1) - (B5). Take $p \in \mathbb{Z}[1,m]$. Define $q = m - p$. Assume N is compact. Let $\tilde{\kappa}$ be another hermitian metric along the fibers of L. If p = 1, define $\chi = 1$. If p > 1 require $c(L,\kappa) \geqq 0$ and $c(L,\tilde{\kappa}) \geqq 0$ and define

$$\chi = \sum_{\nu=0}^{p-1} c(L,\kappa)^{\nu} \wedge c(L,\tilde{\kappa})^{p-1-\nu} .$$

If $p \geqq 1$, define

$$B_p(r) = r^{-2q-2} \int_{M[r]} f^*(\chi) \wedge \upsilon^{q+1} \geqq 0$$

for all $r > 0$. Then B_p increases and is semi-continuous from the right. A constant $\gamma \geqq 0$ exists such that

$$(12.38) \qquad |T_p(r,s,L,\kappa,f) - T_p(r,s,L,\tilde{\kappa},f)| \leqq \gamma\, B_p(r)$$

for all $r > s > 0$.

Remark 1. If $p = 1$, then $B_p(r) = \mathfrak{G}(r)$. If in addition τ is parabolic, then $\mathfrak{G}\ \gamma = \gamma_1$ is constant and

$$(12.39) \qquad |T(r,s,L,\kappa,f) - T(r,s,L,\tilde{\kappa},f)| \leqq \gamma_1$$

for all $r > s > 0$.

Remark 2. If a constant $\gamma_0 \geqq 0$ exists with $0 \leqq c(L,\kappa) \leqq \gamma_0 c(L,\tilde{\kappa})$ on N and if $p > 1$, then a constant $\gamma_2 > 0$ exists such that

$$(12.40) \qquad |T_p(r,s,L,\kappa,f) - T_p(r,s,L,\tilde{\kappa},f)| \leqq \gamma_2\, A_{p-1}(r,L,\tilde{\kappa},f).$$

for all $r > s > 0$.

Remark 3. If $c(L,\tilde{\kappa}) > 0$ on the compact, complex space N, then a constant $\gamma_0 \geqq 0$ exists such that $c(L,\kappa) \leqq \gamma_0\, c(L,\tilde{\kappa})$.

Proof. At first assume that f is holomorphic. By Proposition 10.7, B_p increases with $B_p(0) = \lim_{r\to 0} B_p(r)$ and with

$$(12.41) \qquad B_p(r) = \int_{M(0,r]} f^*(\chi) \wedge \omega^{q+1} + B_p(0)$$

for $r > 0$. If $r \in \mathfrak{C}_\tau$, then

(12.42) $$B_p(r) = \int_{M\langle r\rangle} f^*(\chi) \wedge \sigma_q .$$

A function $g: N \longrightarrow \mathbb{R}$ of class C^∞ exists such that

$$c(L,\kappa) - c(L,\tilde{\kappa}) = dd^c g.$$

Define $0 \leqq \gamma = \operatorname{Max}_{x\in N} |g(x)| < \infty$. Observe

$$c(L,\kappa)^p - c(L,\tilde{\kappa})^p = dd^c g \wedge \chi.$$

Take s and r in $\mathfrak{C}_\tau$ with $0 < s < r$. Then

$$T_p(r,s,L,\kappa,f) - T_p(r,s,L,\tilde{\kappa},f)$$

$$= \int_s^r \int_{M[t]} f^*(dd^c g) \wedge f^*(\chi) \wedge \upsilon^q \, t^{-2q-1} dt$$

$$= \int_s^r \int_{M\langle t\rangle} f^*(d^c g) \wedge f^*(\chi) \wedge \upsilon^q \, t^{-2q-1} dt$$

$$= \frac{1}{2} \int_{M(0,r]} \tau^{-q-1} d\tau \wedge d^c(g\circ f) \wedge f^*(\chi) \wedge \upsilon^q$$

$$= \frac{1}{2} \int_{M(s,r]} \tau^{-q-1} d(g\circ f) \wedge f^*(\chi) \wedge d^c\tau \wedge \upsilon^q$$

$$= \frac{1}{2} \int_{M(s,r]} d(g\circ f) \wedge f^*(\chi) \wedge \sigma_q$$

$$= \frac{1}{2} \int_{M(s,r]} d\left((g\circ f)\, f^*(\chi) \wedge \sigma_q\right) - \frac{1}{2} \int_{M(s,r]} (g\circ f)\, f^*(\chi) \wedge \omega^{q+1}$$

$$= \frac{1}{2} \int_{M\langle r\rangle} (g\circ f)\, f^*(\chi) \wedge \sigma_q - \frac{1}{2} \int_{M\langle s\rangle} (g\circ f)\, f^*(\chi) \wedge \sigma_q$$

$$- \frac{1}{2} \int_{M(s,r]} (g \circ f)\, f^*(\chi) \wedge \omega^{q+1}$$

Clearly $|g \circ f| \leqq \gamma$ on M. Hence (12.41) and (12.42) imply

$$|T_p(r,s,L,\kappa,f) - T_p(r,s,L,\tilde{\kappa},f)|$$

$$\leqq \frac{\gamma}{2} (B_p(r) + B_p(s) + (B_p(r) - B_p(s))) = \gamma\, B(r).$$

Semi-continuity from the right proves the estimate for all $r > s > 0$. If f is meromorphic, go over the graph. The remarks are easily verified; q.e.d.

e) Asymptotic behavior

Assume (B1) - (B5) and (B7). Take $p \in \mathbb{N}[1,m]$. Define $q = m-p$. Then $A_p \geqq 0$ increases. Lemma 11.4 shows

$$(12.43) \qquad 0 \leqq A_p(\infty) = \lim_{r\to\infty} A_p(r) = \lim_{r\to\infty} \frac{T_p(r,s)}{\log r} \leqq \infty.$$

If (B1) - (B5) and (B9) are assumed, then

$$(12.44) \qquad 0 \leqq A_p[\infty] = \lim_{r\to\infty} A_p[r] = \lim_{r\to\infty} \frac{T_p[r,s]}{\log r} \leqq \infty.$$

The proof of the following Lemma is left to the reader (See [82] Lemma 14.14 for a proof.)

Lemma 12.6. Let M and N be complex manifolds of pure dimensions m and n respectively. Assume that M is connected. Let $f: M \longrightarrow N$ be a non-constant holomorphic map. Let S be the set of all $x \in M$ such that the Jacobian of f has rank 0 at x. Then S is thin analytic in M. If ψ and χ are positive, continuous forms of bidegree (1.1) on N and M respectively, then $f^*(\psi) \wedge \chi^{m-1} > 0$ on M - S.

Lemma 12.7. Let M be an irreducible complex space of dimension m.

Let $f: M \longrightarrow N$ be a non-constant holomorphic map into the complex space N. Let ψ and χ be positive, continuous forms of bidegree (1,1) on N and M respectively. Then

$$G = \{x \in \mathfrak{R}(M) \mid f^*(\psi) \wedge \chi^{m-1}(x) > 0\}$$

is open and dense in M.

Proof. W.l.o.g. $M = \mathfrak{R}(M)$. The difficulty of the proof consists in the possibility that $f(M)$ is contained in the singular set of N. Since f is not constant, $p = \operatorname{rank} f > 0$. The set

$$D = \{x \in M \mid \operatorname{rank}_x f < p\}$$

is thin analytic in M. Take $a \in M - D$. Since f has pure rank p on M, an open, connected neighborhood U of a and an open neighborhood W of $f(a)$ exists such that $U \subseteq M - D$ and such that $N' = f(U)$ is a pure p-dimensional analytic subset of W. (Remmert [59] see also [2] Proposition 2.21). Then $f_0 = f: U \longrightarrow N'$ is a surjective holomorphic map of pure rank p. By [2] Lemma 1.27. N' is irreducible. Hence $N_1 = \mathfrak{R}(N')$ is a connected complex manifold of dimension p and $\Sigma(N_1) = N' - N_1$ is thin analytic in N_1. By [2] Lemma 1.25 the set $T = f^{-1}(\Sigma(N_1))$ is thin analytic in U. Hence $U - T = U_1$ is open and connected and $f_1 = f_0: U_1 \longrightarrow N_1$ has pure rank p. Especially f_1 is open and not constant. The set S of all $x \in U_1$ such that the Jacobian rank of f at x is zero is a thin analytic subset of U_1. Let $j_1: N_1 \longrightarrow N$ be the inclusion. Since $p \geqq 1$, the form $j_1^*(\psi) > 0$ is positive. By Lemma 12.6

$$f^*(\psi) \wedge \chi^{m-1} = f_1^*(j_1^*(\psi)) \wedge \chi^{m-1} > 0$$

is positive on $U_1 - S$. Hence $G \supseteq U_1 - S$. Therefore $\overline{G} \supseteq \overline{U}_1 \supseteq U \ni a$, q.e.d.

Theorem 12.8. Assume (B1) - (B5) with $c(L,\kappa) > 0$ on N and $p = 1$. Assume τ is strong. Suppose that $f|B$ is not constant for each branch B of M. Then $A(r) > 0$ if $r > r_\tau^0$ and $A(\infty) > 0$. Also $T(r,s) \longrightarrow \infty$ for $r \longrightarrow \infty$.

Proof. Take $r > r_\tau^0$. Then $M^+(r) \neq \phi$. An open connected subset $U \neq 0$ of $M^+(r)$ exist which is contained in $\Re(M) \cap B$ for some branch B of M. Then $\upsilon > 0$ on U and $f|U$ is not constant. An open, dense subset U_0 of U exists such that

$$f^*(c(L,\kappa)) \wedge \upsilon^{m-1} > 0$$

on $U_0 \subseteq M(r)$. Therefore $A(r) > 0$ by (12.2) for $q = m-1$. Since A increases $A(\infty) > 0$. By (12.42) $T(r,s) \longrightarrow \infty$ for $r \longrightarrow \infty$; q.e.d.

Remark. If only $\upsilon^m \not\equiv 0$ is assume, then $A(r) > 0$ for sufficient large r and $A(\infty) > 0$ and $T(r,s) \longrightarrow \infty$ for $r \longrightarrow \infty$.

Compare the results of this section with Griffiths-King [33] 5(a) (b)(c).

13. The Theorem of Casorati-Weierstrass

The following additional General Assumptions are made

(B10) The meromorphic map f is safe of order p for L.

(B11) The meromorphic map f is almost adapted of order $p-1$ for L.

(B12) The exhaustion τ is strong.

Theorem 13.1. Assume (B1) - (B7) and (B10) - (B12). Take $r > r_\tau^0$. Then $A_p[r] > 0$, and $T_p[r,s] > 0$ if $r > s$. Also $T_p[r,s] \longrightarrow \infty$ for $r \longrightarrow \infty$ and $A_p[\infty] > 0$.

Proof. Define M^+ by (10.17). Then $M^+(r) \neq \phi$ if $r > r_\tau^0$. By Lemma 9.5, an open, dense subset $W \neq 0$ of $M^+(r)$ exists such that

$$f^*(c(L,\ell)^p) \wedge \upsilon^q > 0$$

on W. Hence $A_p[r] > 0$ by (12.7). The rest is a trivial consequence; q.e.d.

Assume (B1) - (B6) and (B10) - (B(12). Define

$$\Gamma_p = \Gamma_p(L,f) = \{a \in G_{p-1}(V) \mid f^{-1}(E_L[a]) \neq \phi\}. \tag{13.1}$$

Lemma 13.2. Γ_p is measurable.

Proof. If f is holomorphic, define $\mathbb{F}_L(f)$ by (5.2) with p replaced by p - 1 and let $\hat{f}: \mathbb{F}_L(f) \longrightarrow G_{p-1}(V)$ be the projection. Then $\mathbb{F}_L(f)$ is analytic. By (5.3) $\hat{f}(\mathbb{F}_L(f)) = \Gamma_p$. Hence Γ_p is measurable. If f is meromorphic, let $\hat{M}$ be the graph of f and let $\vec{f}: \hat{M} \longrightarrow N$ be the projection. Then $\Gamma_p(L,f) = \Gamma_p(L,\vec{f})$ is measurable; q.e.d.

Let I be the integral average operator on $G_{p-1}(V)$. Define $\beta_p(x) =$ 1 if $x \in \Gamma_p$ and $\beta_p(x) = 0$ if $x \notin \Gamma_p$. Then

$$0 \leqq b_p = b_p(L,f) = I(\beta_p) \leqq 1 \tag{13.2}$$

is the measure of Γ_p. Also b_p can be regarded as the probability that $f(M) \cap E_L[a] \neq \phi$.

Theorem 13.3. Assume (B1) - (B6) and (B10) - (B12). Then

$$0 \leqq \theta_p(L,f) = \liminf_{r\to\infty} \frac{A_{p-1}[r]}{T_p[r,s]} \leqq \infty \tag{13.3}$$

does not depend on $s > 0$. Define c(p,k) by (9.1). Then

$$0 \leqq 1 - b_p(L,f) \leqq \tfrac{1}{2}\, c(p,k)\, \theta_p(L,f). \tag{13.4}$$

Especially if $\theta_p(L,f) = 0$, then $f(M) \cap E_L[a] \neq \phi$ for almost all $a \in G_{p-1}(V)$.

Proof. Define β_p as in (13.2). Then $\beta_p(a)N_a(r,s) = N_a r,s)$. Since $\beta_p \leqq 1$, the First Main Theorem implies

$$N_a(r,s) - \beta_p(a)T_p[r,s] \leqq m_a[s] + D_a[r,s]. \tag{13.5}$$

If $r > r_t^0$, then $T_p[r,s] > 0$ by Theorem 13.1. Apply the operator I to (13.5). Then (12.33), (12.34) and (12.35) imply

$$0 \leqq 1 - b_p(L,f) \leqq \tfrac{1}{2} c(p,k) \frac{T_{p-1}[r]}{T_p[r,s]} \tag{13.6}$$

q.e.d.

Theorem 13.4. Assume (B1) - (B6) and (B10) - (B11) with $p = 1$. Let τ be parabolic. Then $f(M) \cap E_L[a] \neq \emptyset$ for almost all $a \in \mathbb{P}(V)$.

Proof. $A_0[r] = \varsigma$ constant. Hence $\theta_1(L,f) = 0$; q.e.d.

Theorem 13.5. Assume (B1) - (B6) and (B10) - (B11) with $p = 1$. Let τ be parabolic. Take $a \in \mathbb{P}(V)$. Assume that f is adapted to a for L. The defect of f for a and L

$$\delta_f^0(a,L) = \liminf_{r\to\infty} \frac{m_a[r]}{T[r,s]} \tag{13.7}$$

is independent of s with

$$0 \leqq 1 - \delta_f^0(a,L) = \limsup_{r\to\infty} \frac{N_a[r,s]}{T[r,s]} \leqq 1. \tag{13.8}$$

Moreover, $\delta_p^0(a,L) = 0$ for almost all $a \in \mathbb{P}(V)$.

Proof. (13.7) and (13.8) follow from (12.31). Fatou's Lemma implies

$$0 \leqq I(\delta_f^0(a,L)) \leqq \liminf_{r\to\infty} I\left(\frac{m_a[r]}{T[r,s]}\right) = \liminf_{r\to\infty} \frac{c(1,k)\,\varsigma}{2T[r,s]} = 0. \quad \text{q.e.d.}$$

If $f(M) \cap E_L[a] = \emptyset$, then $N_a[r,s] = 0$ and $\delta_f^0(a,L) = 1$. Compare the results of this section with Griffiths-King [33] 5c.

V. The Second Main Theorem

14. Jacobian sections

The Carlson-Griffiths-King theory of the Second Main Theorem and the defect relation rests upon the situation on page 195 of [33]. However the considerations on this page are difficult to comprehend and depend on an ad hoc construction. In §14 and §15 Jacobian sections and the Ricci function will be introduced to unravel the mystery and to provide an intrinsic, self contained theory for the Second and Third Main Theorems and the Defect Relations on parabolic manifolds.

Let M and N be complex manifolds of pure dimensions m and n respectively. Let $f: M \longrightarrow N$ be a holomorphic map and let ψ be a positive volume form. Then $\mathrm{Ric}\, f^*(\psi) = f^*(\mathrm{Ric}\, \psi)$ where $f^*(\psi) > 0$ makes sense only if $m = n$. If $m \neq n$, an operator F, called Jacobian section, is introduced which defines a volume from $F[\psi] \geqq 0$ such that $\mathrm{Ric}\, F[\psi] = f^*(\mathrm{Ric}\, \psi)$ where $F[\psi] > 0$.

a) Notations

Let M be a complex manifold of pure dimension $m > 0$. Let T_M be the holomorphic tangent bundle of M. The dual T_M^* is called the holomorphic cotangent bundle. For each holomorphic vector bundle E let $\Gamma(U,E)$ be the vector space of holomorphic sections over the open subset U of M. If $s \in \Gamma(U,E)$, then $Z(s)$ is the zero set of s. Define

$$\Phi_M^p(U) = \Gamma(U, \textstyle\bigwedge_p T_M) \qquad \Omega_M^p(U) = \Gamma(U, \textstyle\bigwedge_p T_M^*)$$

$$\overset{\circ}{\Phi}{}_M^p(U) = \{\chi \in \Phi_M^p(U) \mid Z(\chi) \text{ thin}\} \qquad \overset{\circ}{\Omega}{}_M^p(U) = \{\chi \in \Omega_M^p(U) \mid Z(\chi) \text{ thin}\}$$

$$\overset{*}{\Phi}{}_M^p(U) = \{\chi \in \Phi_m^p(U) \mid Z(\chi) = \phi\} \qquad \overset{*}{\Omega}{}_M^p(U) = \{\chi \in \Omega_M^p(U) \mid Z(\chi) = \phi\}.$$

The elements of $\Phi_M^p(U)$ are the holomorphic vector fields of degree p on U and the elements of $\Omega^p_M(U)$ are the holomorphic forms of degree p on U. Have $K_M = \bigwedge_m T^*_M$ is the <u>canonical</u> bundle and $K^*_M = \bigwedge_m T_M$ is the <u>determinant bundle</u> which is dual to K_M.

Let $\mathfrak{T}(p,m)$ be the set of all increasing injective maps $\mu: \mathbb{N}[1,p] \to \mathbb{N}[1,m]$. If $s_1,\ldots,s_m$ are sections in a vector bundle, define $s_\mu = s_{\mu(1)} \wedge \cdots \wedge s_{\mu(p)}$. If α is a smooth patch on M, then $\alpha = (\alpha_1,\ldots,\alpha_m): U_\alpha \longrightarrow U'_\alpha$ is biholomorphic and $d\alpha_1,\ldots,d\alpha_m$ are holomorphic sections in T^*_M over U_α, in fact a holomorphic frame. The dual frame is denoted by $d^*\alpha_1,\ldots, d^*\alpha_m$. Then $\frac{\partial}{\partial \alpha_j} = d^*\alpha_j$. Also $d\alpha_\mu$ and $d^*\alpha_\mu$ are defined for $\mu \in \mathfrak{T}(p,m)$. Define

$$d\alpha = d\alpha_1 \wedge \cdots \wedge d\alpha_m \qquad d^*\alpha = d^*\alpha_1 \wedge \cdots \wedge d^*\alpha_m .$$

They are holomorphic frames of K_M respectively K^*_M over U.

Let E be a holomorphic vector bundle over the complex space N and let $f: M \longrightarrow N$ be a holomorphic map. Then E_f is the pull back of f and $\tilde{f}: E_f \longrightarrow E$ is defined such that the diagram commutes. The restriction

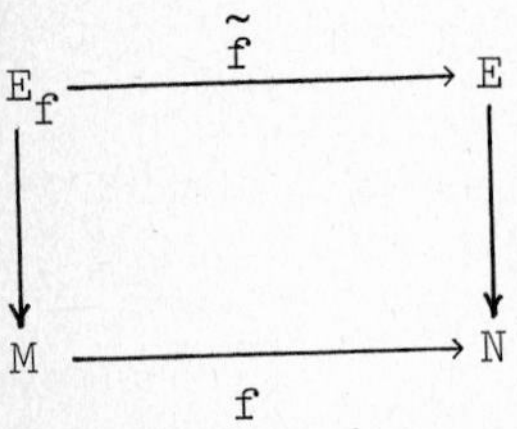

$\tilde{f}_x = \tilde{f}: E_{fx} \longrightarrow E_x$ is an isomorphism. If $s \in \Gamma(U,E)$ and if $\tilde{U} = f^{-1}(U) \neq \phi$, then $s_f \in \Gamma(\tilde{U},E_p)$ is uniquely defined by $s_f(x) = \tilde{f}_x^{-1}(s(f(x)))$ for all $x \in \tilde{U}$. If N is a complex manifold and $s \in \Omega^p_N(U)$, then $s_f \notin \Omega^p_M(\tilde{U})$ and $s_f \neq f^*(s)$. However a homomorphism $\hat{f}: \bigwedge_p T^*_{N,f} \longrightarrow \bigwedge_p T^*_M$ exists uniquely such that $\hat{f}(s_f) = f^*(s)$ for all s.

b) <u>The Jacobian bundle</u>

For the remainder of section 14, M and N will be complex manifolds of pure dimensions m and n respectively and $f: M \longrightarrow N$ will be a holomorphic map. Then $K(f) = K_M \otimes K^*_{Nf}$ is called <u>the Jacobian bundle</u>. A holomorphic section F of K(f) over M is called a <u>Jacobian section</u>. The section F is called <u>effective</u> if Z(F) is thin, its zero divisor θ_F is called the <u>ramification divisor</u> of f for F.

An inner product $(\square,\square) : K^*_N \oplus K_N \longrightarrow \mathbb{C}$ is given and pulls back to $(\square,\square) : K^*_{Nf} \oplus K_{Nf} \longrightarrow \mathbb{C}$ and induces a K_M-valued inner product

$$(\square,\square) : K_f \oplus K_{Nf} = (K_M \otimes K^*_{Nf}) \oplus K_{Nf} \longrightarrow K_M.$$

If U is open in N, define $\tilde{U} = f^{-1}(U)$ throughout this section. If $\tilde{U} \neq \phi$, the Jacobian section F defines a linear map

$$F: \Omega^n_N(U) \longrightarrow \Omega^m_M(\tilde{U}) \quad \text{by} \quad F[\psi] = (F, \psi_f)$$

for all $\psi \in \Omega^n_N(U)$. If $\beta: U_\beta \longrightarrow U'_\beta$ and $\alpha: U_\alpha \longrightarrow U'_\alpha$ are smooth patches on N and M respectively with $U_\beta \subset U$ and $f(U_\alpha) \subseteq U_\beta$, then $\psi = \psi_\beta \, d\beta$ on U_β and $F = F_{\alpha\beta} \, d\alpha \otimes d^*\beta_f$ on U_α, where ψ_β and $F_{\alpha\beta}$ are holomorphic functions. Then $F[\psi] = F_{\alpha\beta} \, (\psi_\beta \circ f) \, d\alpha$ on U_α. The operator F satisfies the following properties.

(P1) If $\psi \in \Omega^n_N(U)$ and $\chi \in \Omega^n_N(U)$, then $F[\psi + \chi] = F[\psi] + F[\chi]$.

(P2) If $\psi \in \Omega^n_N(U)$ and $h \in \Omega^0_N(U)$, then $F[h\,\psi] = (h \circ f)\, F[\psi]$.

(P3) If $V \subseteq U$ are open and $\tilde{V} \neq \phi$, if $\psi \in \Omega^n_N(U)$, then $F[\psi] \mid \tilde{V} = F[\psi | V]$.

A presheaf map $F: \Omega^n_N \longrightarrow \Omega^m_M$ satisfying (P1) - (P3) is called a <u>holomorphic volume homomorphism.</u>

(P4) A holomorphic volume homomorphism $F: \Omega^n_N \longrightarrow \Omega^m_M$ defines uniquely a Jacobian section $\tilde{F}$ such that $F[\psi] = \tilde{F}[\psi]$.

(P5) If $\psi \in \overset{*}{\Omega}{}^n_N(U)$ and $\tilde{U} \neq \phi$, then $Z(F) \cap \tilde{U} = Z(F[\psi])$. If F is effective, the divisors of F and $F[\psi]$ agree: $\theta_F | \tilde{U} = \theta_{F[\psi]}$.

These simple facts are easily proved, see also [82] §14c).

c) <u>The existence of Jacobian sections</u>

If M is Stein, effective holomorphic sections exist by general theory. However, more constructive methods shall be investigated.

At first assume $q = m - n \geqq 0$. Then one and only one holomorphic section Df of $(\underset{n}{\Lambda} T^*_M) \otimes K^*_{Nf}$ exists such that $Df|\tilde{U} = f^*(\psi) \otimes \psi^*_f$ whenever U is open in N with $\tilde{U} \neq \emptyset$ and $\psi \in \tilde{\Omega}^{*n}_N(U)$. Here ψ^* is the dual frame to ψ. If $\alpha \in \mathfrak{S}_M$ and $\beta \in \mathfrak{S}_N$ are smooth patches with $f(U_\alpha) \subseteq U_\beta$, then

$$f^*(d\beta) = \sum_{\mu \in \mathfrak{T}(n,m)} \Delta_\mu \, d\alpha_\mu$$

$$Df|\tilde{U}_\alpha = \sum_{\mu \in \mathfrak{T}(n,m)} \Delta_\mu \, d\alpha_\mu \otimes d^*\beta_f.$$

Hence Z(Df) is the set of all $x \in M$ such that the rank of the Jacobian matrix at x is smaller than n. ([82] Proposition 17.2). The section Df is called the differential of f. If $N = \mathbb{C}$, then it is the exterior derivative of f. If $n = m$, then $\underset{n}{\Lambda} T^*_M = K_M$ and Df is a Jacobian section which is effective if and only if f has strict rank $n = m$. In this case the divisor θ_{Df} of Df is called the branching divisor of f.

Assume $q = m - n \geqq 0$ and let φ be a holomorphic form of degree q on M. The Jacobian section $F_\varphi = \varphi \wedge Df$ is called the section induced by φ and φ is said to be effective for f if F_φ is effective. (Observe the wedge product extends from $\underset{q}{\Lambda} T^*_M$, $\underset{n}{\Lambda} T^*_M$ to $\underset{q}{\Lambda} T^*_M$, $\underset{n}{\Lambda} T^*_M \otimes K_{Nf}$ and becomes K(f) valued). Equivalently, F_φ can be described by its action

$$F_\varphi[\chi] = \varphi \wedge f^*(\chi)$$

whenever $\chi \in \Omega^n_N(U)$ and U open in N with $\tilde{U} \neq \emptyset$. (See [82] Lemma 17.3). Trivially $Z(Df) \subseteq Z(F_\varphi)$. Hence a necessary condition for φ to be effective is that f has strict rank n.

The complex manifold M of pure dimension m is said to have holomorphic rank p, if there exist p analytically independent holomorphic functions; i.e. $\alpha_1, \ldots, \alpha_p$ holomorphic exist such that $d\alpha_1 \wedge \ldots \wedge d\alpha_p \not\equiv 0$ on each connectivity component of M; equivalently, a holomorphic map $\alpha: M \longrightarrow \mathbb{C}^p$ of strict rank p exists. (See [2]). If M is Stein, M has holomorphic rank m.

Theorem 14.1. Assume M has holomorphic rank m. Assume $m - n = q \geqq 0$. Then there exists a holomorphic form φ of degree q on M which is effective for f if and only if f has strict rank n.

Proof. W.ℓ.o.g. M is assumed to be connected. Holomorphic functions $\alpha_\mu : M \longrightarrow \mathbb{C}$ and a point $a \in M$ exists such that $(d\alpha_1 \wedge \ldots \wedge d\alpha_m)(a) \neq 0$. Take a smooth patch $\beta: U_\beta \longrightarrow U'_\beta$ on N with $f(a) \in U_\beta$. An open connected neighborhood U_α of a with $f(U_\alpha) \subset U_\beta$ exists such that $\alpha = (\alpha_1, \ldots, \alpha_m)$ defines a smooth patch $\alpha: U_\alpha \longrightarrow U'_\alpha$. Then

$$f^*(d\beta) = \sum_{\mu \in \mathfrak{T}(n,m)} \Delta_\mu \, d\alpha_\mu .$$

Assume that f has strict rank n on M. Then $\iota \in \mathfrak{T}(n,m)$ exists such that $\Delta_\iota \not\equiv 0$ on U_α. W.ℓ.o.g assume that $\iota: \mathbb{N}[1,n] \longrightarrow \mathbb{N}[1,m]$ is the inclusion. Define

$$\varphi = d\alpha_{n+1} \wedge \ldots \wedge d\alpha_m \in \Omega^q_M(M).$$

Then $F_\varphi[d\beta] = \varphi \wedge f^*(d\beta) = (-1)^{nq} \Delta_\iota \, d\alpha$. Therefore

$$Z(F_\varphi) \cap U_\alpha = Z(F_\varphi[d\beta]) \cap U_\alpha = \Delta_\iota^{-1}(0).$$

Thus $Z(F_\varphi)$ is thin in U_α hence in M; q.e.d.

The case $m < n$ is more difficult. A holomorphic section $\varphi \in \Gamma(M, \underset{q}{\Lambda} T_{Nf})$ is called a holomorphic field on f over M of degree q. If $q = n - m > 0$, then φ induces a Jacobian section F_φ. Recall the homomorphism $\hat{f}: K_{N,f} \longrightarrow \underset{n}{\Lambda} T^*_M$ mentioned in a) such that $\hat{f}(\psi_f) = f^*(\psi)$ for all $\psi \in \Omega^n_N(U)$.

Also the interior product

$$\llcorner : \underset{n}{\Lambda} T^*_N \oplus \underset{q}{\Lambda} T^*_N \longrightarrow \underset{m}{\Lambda} T^*_M$$

pulls back to an interior product

$$\llcorner : \underset{n}{\Lambda} T^*_{Nf} \oplus \underset{q}{\Lambda} T^*_{N,f} \to \underset{m}{\Lambda} T^*_{M,f}.$$

Then F_φ is defined uniquely by the following property

(P) Let $\chi \in \overset{*}{\Omega}{}^n_N(U)$ be a holomorphic frame of K_N over the open subset U of N with $\tilde{U} = f^{-1}(U) \neq \phi$. Let χ^* be the dual frame. Then

$$F_\varphi | \tilde{U} = \hat{f}(\chi_f \wedge \varphi) \otimes \chi^*_f.$$

Equivalently F_φ can be defined by its action

(Q) Let U be open in N with $\tilde{U} \neq \phi$. Take $\psi \in \Omega^n_N(U)$. Then $F_\varphi[\psi] = \hat{f}(\psi_f \llcorner \varphi)$.

The proof requires only elementary calculations. For details see [82] Lemma 17.5.

Again $\varphi \in \Gamma(M, \underset{q}{\Lambda} T^*_{Nf})$ is said to be <u>effective</u> for f if F_φ is effective.

<u>Theorem 14.2.</u> Assume M is Stein and $n - m = q > 0$. Then there exists a holomorphic field φ on f over M of degree q such that φ is effective for f if and only if f has strict rank m.

<u>Proof.</u> W.ℓ.o.g. M is assumed to be connected. <u>Assume an effective φ is given.</u> Let S be the set of all $x \in M$ such that the rank of the Jacobian of f at x is smaller than m. Take $a \in M$. Take smooth patches $\alpha: U_\alpha \longrightarrow U'_\alpha$ on M with $a \in U_\alpha$ and $\beta: U_\beta \longrightarrow U'_\beta$ on N with $f(U_\alpha) \subseteq U_\beta$. Then

$$\varphi = \sum_{\mu \in \mathfrak{T}(q,n)} \varphi_\mu d^*\beta_{\mu f} \tag{14.1}$$

on U_α. Also $f^*(d\beta_\nu) = \Delta_\nu d\alpha$ for all $\nu \in \mathfrak{T}(m,n)$. Then

$$S \cap U_\alpha = \bigcap_{\nu \in \mathfrak{T}(m,n)} \Delta_\nu^{-1}(0).$$

If $\mu \in \mathfrak{T}(q,n)$, then $\mu^\perp \in \mathfrak{T}(m,n)$ is uniquely defined such that $(\mu,\mu^\perp)$ is a permutation of $\mathbb{N}[1,m]$. Then $\perp: \mathfrak{T}(q,n) \longrightarrow \mathfrak{T}(m,n)$ is bijective. Observe $d\beta = d\beta_1 \wedge \dots \wedge d\beta_m$. Also

$$d\beta_f \llcorner \varphi = \sum_{\mu\in\mathfrak{T}(q,n)} \operatorname{sign}(\mu^\perp,\mu)\varphi_\mu \, d\beta_{\mu^\perp f}.$$

Hence

$$(14.2) \qquad F_\varphi[d\beta_f] = \hat{f}(d\beta_f \llcorner \varphi) = \sum_{\mu\in\mathfrak{T}(q,n)} \operatorname{sign}(\mu^\perp,\mu)\varphi_\mu \Delta_{\mu^\perp}\, d\alpha.$$

Define $\psi = \sum_{\mu\in\mathfrak{T}(q,n)} \operatorname{sign}(\mu^\perp,\mu)\varphi_\mu \Delta_{\mu^\perp}$. Then

$$Z(F_\varphi) \cap U_\alpha = Z(F_\varphi[d\beta_f]) \cap U_\alpha = \psi^{-1}(0) \supseteq S \cap U_\alpha.$$

Because $Z(F_\varphi)$ is thin, also $S \cap U_\alpha$ is thin. Since M is connected, S is thin. Hence f has (strict) rank m.

Assume f has strict rank m on M. Take S, a, α, β, Δ_ν as above. Since S is thin $a \in M - S$ can be chosen. Then $\iota \in \mathfrak{T}(m,n)$ exists such that $\Delta_\iota(a) \neq 0$. Define $j = \iota^\perp \in \mathfrak{T}(q,n)$. Because M is Stein, T_{Nf} is ample. Hence global holomorphic sections $\gamma_\lambda \in \Gamma(M,T_{Nf})$ exist such that $\gamma_\lambda(a) = d^*\beta_{\lambda f}(a)$ for $\lambda = 1,\dots,n$. Define $\varphi = \lambda_j \in \Gamma(M,\Lambda_q T_{N,f})$. Then $\varphi(a) = \gamma_j(a) = d^*\beta_{jf}(a)$. Hence (14.1) holds with $\varphi_j(a) = 1$ and $\varphi_\mu(a) = 0$ if $j \neq \mu \in \mathfrak{T}(q,n)$. Also (14.2) holds with

$$F_\varphi[d\beta_f](a) = \operatorname{sign}(\iota,j)\,\Delta_\iota(a) \neq 0.$$

Observe $Z(F_\varphi[d\beta_f]) \cap U_\alpha = Z(F_\varphi) \cap U_\alpha$. Hence $a \in M - Z(F_\varphi)$. The analytic set $Z(F_\varphi)$ is thin. Hence φ is effective. q.e.d.

(The proof of the corresponding Theorem 17.6 in [82] is false).

If $m = n$, then $F_1[\quad] = f^*$ is the pull back of forms by either construction and $Z(F_1)$ is the zero set of the Jacobian determinant of f.

d) Jacobian sections as operators on volume forms

Let $A_k^p(U)$ (respectively $A_k^{p,q}(U)$) be the vector space of forms of class C^k and degree p (respectively bidegree (p,q)) on M. Define

(14.3) $$i_p = \left(\frac{i}{2\pi}\right)^p (-1)^{\frac{p(p-1)}{2}} p!$$

Then a Jacobian section $F \in \Gamma(M,K(f))$ operates also on forms of degree 2n. Let U be open in N with $\tilde{U} = f^{-1}(U) \neq \phi$. A linear map $F: A_k^{2n}(U) \longrightarrow A_k^{2m}(\tilde{U})$ is uniquely defined as follows: Take $\psi \in A_k^{2n}(U)$. Let $\beta: U_\beta \longrightarrow U'_\beta$ be a smooth patch on N with $U_\beta \subseteq U$. Then $\psi | U_\beta = i_n \psi_\beta d\beta \wedge d\overline{\beta}$ and

(14.4) $$F[\psi] \mid \tilde{U}_\beta = i_m(\psi \circ f)\, F[d\beta] \wedge \overline{F[d\beta]}.$$

Clearly, this extended operation of F satisfies the following properties

(Q1) If $\psi \in A_k^{2n}(U)$ and $\chi \in A_k^{2n}(U)$, then $F[\psi + \chi] = F[\psi] + F[\chi]$.

(Q2) If $\psi \in A_k^{2n}(U)$ and $h \in A_k^0(U)$, then $F[h\psi] = (h \circ f)\, F[\psi]$.

(Q3) If $0 \leqq \psi \in A_k^{2n}(U)$, then $F[\psi] \geqq 0$.

(Q4) If $0 < \psi \in A_k^{2n}(U)$, then $F[\psi] > 0$ on $\tilde{U} - Z(F)$.

(Q5) If $V \subseteq U$ are open with $\tilde{V} \neq \phi$, if $\psi \in A_k^{2n}(U)$, then $F[\psi] | \tilde{V} = F[\psi | V]$.

(Q6) If $\varphi \in \Omega_N^n(U)$ and $\psi \in \Omega_N^n(U)$ with $\tilde{U} \neq \phi$, then

(14.5) $$F[i_n\, \varphi \wedge \overline{\psi}] = i_m\, F[\varphi] \wedge \overline{F[\psi]}.$$

(Q7) If $q = m - n \geqq 0$, if $\varphi \in \Omega_M^q(M)$ and $\psi \in A_K^{2n}(U)$, then

(14.6) $$F_\varphi[\psi] = \binom{m}{q}\, i_q\, \varphi \wedge \overline{\varphi} \wedge f^*(\psi).$$

For more details see [82] §17 d). In fact F can be factored out of $F[\psi]$:

Proposition 14.3. Let κ be a hermitian metric along the fibers of $K(f)$. Let F be an effective Jacobian section. Let U be open in N with $\tilde{U} = f^{-1}(U) \neq \phi$. Take $0 < \psi \in A_k^{2n}(U)$. Then there exists one and only one $0 < \hat{\psi} \in A_k^{2m}(\tilde{U})$ such that $F[\psi] = |F|_k^2 \hat{\psi}$ on $\tilde{U}$.

Proof. Since $\hat{F}[\psi] > 0$ and $|F|_k^2 > 0$ on $\tilde{U} - Z(F)$, the form $\hat{\psi} > 0$ is well defined, of class C^k and of degree $2m$ on $\tilde{U} - Z(F)$. So $\hat{\psi}$ has to be continued over the thin analytic set $Z(F) \cap \tilde{U}$.

Take $a \in Z(F) \cap \tilde{U}$. Smooth patches $\alpha: U_\alpha \longrightarrow U'_\alpha$ on M and $\beta: U_\beta \longrightarrow U'_\beta$ on N exist with $a \in U_\alpha$ and $f(U_\alpha) \subseteq U_\beta \subseteq U$. Then $F = g\, d\alpha \otimes d^*\beta_f$ on U_α where g is holomorphic on U_α with $Z(F) = g^{-1}(0)$. Also $1/h = |d\alpha \otimes d^*\beta_f|_\kappa^2 > 0$ is a function of class C^∞ on U_α. Then $h|F|_\kappa^2 = |g|^2$ on U_α. Also $F[d\beta] = g\, d\alpha$ on U_α. A function $\psi_\beta > 0$ of class C^k exists on U_β such that $\psi = i_n\, \psi_\beta\, d\beta \wedge d\bar{\beta}$ on U_β. Therefore

$$F[\psi]|U_\alpha = i_m(\psi_\beta \circ f)\; F[d\beta] \wedge \overline{F[d\beta]}$$

$$= (\psi_\beta \circ f)\; |g|^2\; i_m d\alpha \wedge d\alpha$$

$$= |F|_\kappa^2\; (\psi_\beta \circ f)\; h\; i_m d\alpha \wedge d\alpha$$

where $(\psi_\beta \circ f)\; h\; i_m\; d\alpha \wedge d\bar{\alpha}$ is positive and of class C^k on U_α and continuous $\hat{\psi}$ from $U_\alpha - Z(F)$ onto U_α; q.e.d.

Theorem 14.4. Let F be an effective Jacobian section for f. Let U be open in N with $\tilde{U} = f^{-1}(U) \neq \phi$. Let $\psi > 0$ be a positive form of class C^2 and of degree $2m$ on U. Then $F[\psi] > 0$ on $\tilde{U} - Z(F)$ and $\mathrm{Ric}\, F\,[\psi] = f^*(\mathrm{Ric}\,\psi)$ on $\tilde{U} - Z(F)$. (Observe that $f^*(\mathrm{Ric}\,\psi)$ does not depend on F.)

Proof. Take $a \in \tilde{U} - Z(F)$. Smooth patches $\alpha: U_\alpha \longrightarrow U'_\alpha$ on M and $\beta: U_\beta \longrightarrow U'_\beta$ on N exist such that $a \in U_\alpha \subseteq \tilde{U} - Z(F)$ and $f(U_\alpha) \subseteq U_\beta \subseteq U$. Then a holomorphic function g on U_α exists such that $F = g\, d\alpha \wedge d^*\beta_f$ with $g(x) \neq 0$ for all $x \in U_\alpha$. Then $F[d\beta] = g\, d\alpha$ on U_α. A positive function $\psi_\beta > 0$ of class C^2 exists on U_β such that $\psi = i_m \psi_\beta \, d\beta \wedge d\bar\beta$ on U_β. Then

$$F[\psi] = (\psi \circ f) \, |g|^2 i_m d\alpha \wedge d\bar\alpha \qquad \text{on } U_\alpha$$

$$\text{Ric } F[\psi] = dd^c(\log (\psi_\beta \circ f) \, |g|^2) = f^*(dd^c \log \psi_\beta) = f^*(\text{Ric } (\psi)).$$

q.e.d.

15. The Ricci function

In this section M is a complex manifold of pure dimension m and τ is an exhaustion of M. Let Ω be a positive form of class C^2 and degree 2m on an open subset U of M such that M - U has measure zero. If $\text{Ric } \Omega \wedge \upsilon^{m-1}$ is locally integrable, define

$$\text{(15.1)} \qquad \text{ric } (r,\Omega) = r^{2-2m} \int_{M[r]} \text{Ric } \Omega \wedge \upsilon^{m-1} \qquad \text{for } r > 0$$

$$\text{(15.2)} \qquad \text{Ric } (r,s,\Omega) = \int_s^r \text{ric } (t,\Omega) \frac{dt}{t} \qquad \text{for } 0 < s < r$$

as the <u>Ricci function</u> of Ω. If Ω is of class C^∞ on M, then ric (r,Ω) is the spherical image and Ric (r,s,Ω) the characteristic function for (K_M, κ_Ω) and the identity map. Since υ^m may have too many zeros, (15.1) and (15.2) may not apply directly to υ^m. However an indirect way exists

<u>Lemma 15.1.</u> Let τ be a logarithmic pseudo-convex exhaustion of M. Take a continuous, positive form Ω of degree 2m on M. Define v by $\upsilon^m = v \, \Omega$ on M. Define $\log v(x) \, \upsilon^m(x) = 0$ if $\upsilon^m(x) = 0$. Then $(\log v) \, \upsilon^m$ is locally integrable over M. Define $\mathfrak{C}_\tau^0 = \{r \in \mathfrak{C}_\tau \mid \int_{M\langle r\rangle} (\log v)\sigma \text{ exists}\}$ Then $\mathfrak{C}_\tau^0$ does not depend on the choice of Ω and $\mathbb{R}_+ - \mathfrak{C}_\tau^0$ has measure zero

<u>Proof.</u> If $\tilde\Omega$ is another choice of Ω determining $\tilde v$. Then $\Omega = h \, \tilde\Omega$ where $h: M \longrightarrow \mathbb{R}^+$ is continuous. Then $\tilde v = h \, v$ and $\log \tilde v = \log h + \log v$ where $\int_{M\langle r\rangle} \log h \, \sigma$ exists for all $r \in \mathfrak{C}_\tau$. Hence $\mathfrak{C}_\tau^0$ does not depend on Ω.

Take $r > 0$. Since v is continuous, a constant K exists such that $0 \leqq v \leqq K$ on $M[r]$. Then

$$0 \leqq \int_{M[r]} \log \frac{K}{v} \; \upsilon^m \leqq 2 \int_{M[r]} \left(\frac{K}{v}\right)^{\frac{1}{2}} v \,\Omega \leqq 2K \int_{M[r]} \Omega < \infty.$$

Hence $(\log v)\,\upsilon^m$ is locally integrable on M. Since $\omega^m \geqq 0$, (10.6) implies $0 \leqq m\,\tau^{m-1}\,d\tau \wedge \sigma \leqq \upsilon^m$. Hence Lemma 10.2 gives

$$\infty > \int_{M[r]} (\log v)\; m\tau^{m-1} d\tau \wedge \sigma = 2m \int_0^r \Big(\int_{M\langle t\rangle} \log v \;\sigma\Big) t^{2m-1} dt.$$

Therefore $\mathbb{R}_+ - \mathfrak{C}^0_\tau$ has measure zero; q.e.d.

Assume τ is a parabolic exhaustion of M. Take a positive form Ω of degree $2m$ and class C^2 on M. Define $v\colon M \longrightarrow \mathbb{R}_+$ by $\upsilon^m = v\Omega$. For $0 < s < r$ with $s \in \mathfrak{C}^0_\tau$ and $r \in \mathfrak{C}^0_\tau$ the <u>Ricci function of τ</u> is defined by

$$(15.3) \quad \mathrm{Ric}_\tau(r,s) = \frac{1}{2} \int_{M\langle r\rangle} \log v \;\sigma - \frac{1}{2} \int_{M\langle s\rangle} \log v \;\sigma + \mathrm{Ric}(r,s,\Omega).$$

<u>Lemma 15.2.</u> The Ricci function Ric_τ of τ does not depend on the choice of Ω.

<u>Proof.</u> Let $\tilde{\Omega}$ be another choice defining $\tilde{v}$. Then $\Omega = h\,\tilde{\Omega}$ where h is a positive function of class C^2 on M with $\tilde{v} = h\,v$.

$$\frac{1}{2} \int_{M\langle r\rangle} \log v \;\sigma - \frac{1}{2} \int_{M\langle s\rangle} \log v \;\sigma + \mathrm{Ric}(r,s,\Omega)$$

$$- \frac{1}{2} \int_{M\langle r\rangle} \log \tilde{v} \;\sigma + \frac{1}{2} \int_{M\langle s\rangle} \log \tilde{v} \;\sigma - \mathrm{Ric}(r,s,\tilde{\Omega})$$

$$= \frac{1}{2}\int_{M\langle r\rangle} \log h\,\sigma - \frac{1}{2}\int_{M\langle s\rangle} \log h\,\sigma - \int_s^r \int_{M[t]} dd^c \log h \wedge \upsilon^{m-1} \frac{dt}{t^{2m-1}}$$

$$= \frac{1}{2}\int_{M(s,r]} d\log h \wedge \sigma - \int_s^r \int_{M\langle t\rangle} d^c \log h \wedge \upsilon^{m-1} \frac{dt}{t^{2m-1}}$$

$$= \frac{1}{2}\int_{M(s,r]} d\log \tau \wedge d^c \log h \wedge \omega^{m-1} - \int_s^r \int_{M\langle t\rangle} d^c \log h \wedge \omega^{m-1} \frac{dt}{t}$$

$$= \frac{1}{2}\int_s^r \Big(\int_{M\langle t\rangle} d^c \log h \wedge \omega^{m-1}\Big) d\log t^2 - \int_s^r \int_{M\langle t\rangle} d^c \log h \wedge \omega^{m-1} \frac{dt}{t}$$

$$= 0$$

q.e.d.

If τ is a parabolic exhaustion with $\upsilon > 0$ on M, then $\mathrm{Ric}_\tau(r,s) = \mathrm{Ric}(r,s,\upsilon^m)$ extends to all $r > s > 0$. However, the only known example of a parabolic manifold with $\upsilon > 0$ is $\mathbb{C}^m$ with $\tau(\mathfrak{z}) = |\mathfrak{z}|^2$. There are other cases, in which Ric_τ can be computed.

<u>Lemma 15.3.</u> Let τ be a parabolic exhaustion of M. Let ζ be a holomorphic form of degree m such that Z(s) is thin. Let θ_ζ be the (zero) divisor of ζ. On M^+, define a function $Z \geqq 0$ of class C^∞ by $Z\upsilon^m = i_m \zeta \wedge \overline{\zeta}$. Set $Z = 1$ on $M - M^+$. If $r \in \mathfrak{C}_\tau$, then $r \in \mathfrak{C}_\tau^0$ if and only if $(\log Z)\sigma$ is integrable over $M\langle r\rangle$. If $s \in \mathfrak{C}_\tau^0$ and $r \in \mathfrak{C}_\tau^0$, then

$$(15.4) \qquad \mathrm{Ric}_\tau(r,s) + \frac{1}{2}\int_{M\langle r\rangle} \log Z\,\sigma - \frac{1}{2}\int_{M\langle s\rangle} \log Z\,\sigma = N_{\theta_\zeta}(r,s).$$

<u>Proof.</u> Take a form $\Omega > 0$ of class C^∞ and degree 2m on M. Define v and u by $\upsilon^m = v\Omega$ and $i_m \zeta \wedge \zeta = u\Omega$. Then $Zv = u$ on M^+. Let $\kappa = \kappa_\Omega$ be the hermitian metric defined by Ω along the fibers of K_M. Let $f\colon M \longrightarrow M$ be the identity. Then $T_f(r,s,K_M,\kappa,f) = \mathrm{Ric}(r,s,\Omega)$. Take a finite dimensional linear subspace V of $\Gamma(M,K_n)$ with $\zeta \in V$. Take a hermitian metric ℓ on V with $|\zeta| = 1$. Let $\eta\colon M \times V \longrightarrow K_M$ be the

evaluation map $\eta(x,s) = s(x)$. Then η is a semi-amplification since $Z(s)$ is thin. Define $a = \mathbb{P}(\zeta) \in \mathbb{P}(V)$. Then $Z(\zeta) = E_{K_M}[a]$ and $\theta_f^a[K_M] = \theta_\zeta$. Also f is adapted to a for K_M. Hence $N_a(r,s,K_M,f) = N_{\theta_\zeta}(r,s)$. By (2.2) and Lemma 4.2 $u = |\zeta|_\kappa^2 = \|a,\square\|_\kappa^2$. Hence

$$\frac{1}{2}\int_{M\langle r\rangle} \log u\ \sigma = -\ m_a(r,K_N,\kappa,f) \qquad \text{if } r \in \mathfrak{E}_\tau.$$

If $0 < s < r \in \mathfrak{E}_\tau$ and $s \in \mathfrak{E}_\tau$, the First Main Theorem (12.28) implies

$$(15.5)\quad \operatorname{Ric}(r,s,\Omega) + \frac{1}{2}\int_{M\langle r\rangle} \log u\ \sigma - \frac{1}{2}\int_{M\langle s\rangle} \log u\ \sigma = N_{\theta_\zeta}(r,s).$$

Now $Zv = u$, (15.3) and (15.5) show (15.4); q.e.d.

Especially, if $i_m\zeta \wedge \overline{\zeta} = \upsilon^m$, then $Z \equiv 1$ and $\operatorname{Ric}_\tau(r,s) = N_{\theta_\zeta}(r,s)$.

<u>Theorem 15.4.</u> Let $\beta = M \longrightarrow \mathbb{C}^m$ be a proper, holomorphic map of strict rank m. Let $\nu = \theta_{D_\beta}$ be the branching divisor of β. Then $\tau = |\beta|^2$ is a parabolic exhaustion of M with $\mathfrak{E}_\tau = \mathfrak{E}_\tau^0 = \mathbb{R}^+$ and $\operatorname{Ric}_\tau(r,s) = N_\nu(r,s)$ for all $0 < s < r \in \mathbb{R}^+$.

<u>Proof.</u> Let $\alpha = (\alpha_1,\ldots,\alpha_m): \mathbb{C}^m \longrightarrow \mathbb{C}^m$ be the identity. Then $D\beta = \beta^*(d\alpha) \otimes d^*\alpha_\beta$. Hence $\zeta = \beta^*(d\ \alpha) \in \Omega_M^m(M)$ and $\theta_\zeta = \theta_{D\beta} = \nu$. Also $i_m\zeta \wedge \zeta = \upsilon^m$. Therefore $Z = 1$ and Lemma 15.3 completes the proof; q.e.d.

Let M and N be complex manifolds of dimensions m and n respectively. Let $f: M \longrightarrow N$ be a holomorphic map. Let N_0 be open in M such that $M - f^{-1}(N_0)$ has measure zero in M. Let τ be an exhaustion of M. Let ψ be a positive form of class C^2 and degree 2n on N_0 such that

(15.6) $$\mathrm{ric}(r,\psi,f) = r^{2-2m}\int_{M[r]} f^*(\mathrm{Ric}\,\psi)\wedge \upsilon^{m-1}$$

exists for all $r > 0$. For $0 < s < r$ the <u>Ricci function</u> of ψ for f is defined by

(15.7) $$\mathrm{Ric}(r,s,\psi,f) = \int_s^r \mathrm{ric}(t,\psi,f)\,\frac{dt}{t}.$$

<u>Theorem 15.5</u>. Let M and N be complex manifolds of dimensions m and n respectively. Let τ be a parabolic exhaustion of M. Let $f: M \longrightarrow N$ be holomorphic map. Let F be an effective Jacobian section of f. Let ψ be a positive volume form of class C^∞ and degree 2n on N. A function $v \geqq 0$ of class C^∞ on $M - Z(F)$ is defined by $\upsilon^m = vF[\psi]$. If $0 < s < r \in \mathfrak{C}^0_\tau$ and $s \in \mathfrak{C}^0_\tau$, then

(15.8) $$\mathrm{Ric}_\tau(r,s) = N_{\theta_F}(r,s) + \mathrm{Ric}(r,s,\psi,f)$$
$$+ \frac{1}{2}\int_{M\langle r\rangle} (\log v)\sigma - \frac{1}{2}\int_{M\langle s\rangle} (\log v)\sigma.$$

<u>Proof.</u> Take a hermitian metric κ along the fibers of K(f). By Proposition 14.3 a form $\hat{\psi} > 0$ of class C^∞ and degree 2m on M exists such that $F[\psi] = |F|^2_\kappa\,\hat{\psi}$. Hence $\upsilon^m = vF[\psi] = v|F|^2_\kappa\,\hat{\psi}$. If $0 < s < r \in \mathfrak{C}^0_\tau$ and $s \in \mathfrak{C}^0_\tau$, then

$$\mathrm{Ric}_\tau(r,s) = \frac{1}{2}\int_{M\langle r\rangle} (\log v)\sigma - \frac{1}{2}\int_{M\langle s\rangle} \log v\ \sigma$$

(15.9) $$+ \frac{1}{2}\int_{M\langle r\rangle} \log|F|^2_\kappa\,\sigma - \frac{1}{2}\int_{M\langle s\rangle} \log|F|^2_\kappa\,\sigma$$
$$+ \mathrm{Ric}(r,s,\hat{\psi}).$$

Theorem 14.4 implies

$$f^*(\mathrm{Ric}\,\psi) = \mathrm{Ric}\,F[\psi] = \mathrm{Ric}\,\hat{\psi} + dd^c\log|F|_\kappa^2 = \mathrm{Ric}\,\hat{\psi} - c(K(f),\kappa).$$

If $\iota: M \longrightarrow M$ is the identity, then

$$(15.10) \qquad \mathrm{Ric}(r,s,\hat{\psi}) = \mathrm{Ric}(r,s,\psi,f) + T(r,s,K(f),\kappa,\iota).$$

Take a finite dimensional linear subspace V of $\Gamma(M,K(f))$ with $F \in V$. Observe $F \not\equiv 0$. Take a hermitian metric ℓ on V with $|F| = 1$. The evaluation map $\eta: M \times V \longrightarrow K(f)$ defined by $\eta(x,s) = s(x)$ is a semi-amplification. Define $a = \mathbb{P}(F) \in \mathbb{P}(V)$. Then $Z(F) = E_{K(f)}[a]$ and $\theta_L^a[K[f]] = \theta_F$. Hence $N_a(r,s,K(f),\iota) = N_{\theta_F}(r,s)$. If $x \in M$, then $\|x,a\|_\kappa = |F(x)|_\kappa$. Hence

$$m_a(r,K(f),\kappa,\iota) = -\int_{M\langle r\rangle} \log|F|_\kappa \sigma.$$

If $0 < s < r \in \mathfrak{S}_\tau$ and $s \in \mathfrak{S}_\tau$ the First Main Theorem (12.28) shows

$$(15.11) \quad T(r,s,K(f),\kappa,\iota) = N_{\theta_F}(r,s) - \int_{M\langle r\rangle} \log|F|_\kappa \sigma + \int_{M\langle s\rangle} \log|F|_\kappa \sigma.$$

Now (15.9), (15.10) and (15.11) imply (15.8); q.e.d.

The Ricci function and the Jacobian section clarify the situation tremendously. They do not appear in [33], but Theorem 15.5 can be considered an interpretation of Lemma 6.20 in Griffiths-King [33]. The idea to define the Ricci function by (15.3) originated in [67] Satz 14.1.

16. The singular volume form of Carlson and Griffiths

a) General position

Let N be a complex manifold of pure dimension $n > 0$. Since N is a manifold, divisors are identified with their multiplicity functions. Divisors $\nu_1,\ldots,\nu_q$ are said to be in general position (i.e. have normal crossings) if the following condition holds.

(G) Take $x \in N$. Then indices $j_1, \ldots j_p$ and holomorphic functions $f_1, \ldots, f_p$ exist on an open neighborhood U of x such that $f_\mu(x) = 0$ and $\theta_{f_\mu} = \nu_{j_\mu} | U$ for all $\mu = 1, \ldots, p$ and such that $U \cap \operatorname{supp} \nu_\mu = \emptyset$ if $\mu \neq j_\lambda$ for $\lambda = 1, \ldots, p$. Also

$$(16.1) \qquad (df_1 \wedge \cdots \wedge df_p)(z) \neq 0 \qquad \text{for all } z \in U.$$

If $\nu_1, \ldots, \nu_q$ are in general position, each divisor ν_μ is holomorphic and $A_\mu = \operatorname{supp} \nu_\mu$ has no singular points. Also $\nu_\mu | A_\mu = 1$ for $\mu = 1, \ldots, q$. Also

$$0 \leqq p = \# \{\mu \in \mathbb{N}[1,q] \mid x \in \operatorname{supp} \nu_\mu\} \leqq n.$$

Let $L_1, \ldots, L_q$ be holomorphic line bundles on N. Let $s_1, \ldots, s_q$ be holomorphic sections of $L_1, \ldots, L_q$ respectively. Then $s_1, \ldots, s_q$ are said to be in <u>general position</u> if each $Z(s_\lambda)$ is thin in N and if $\theta_{s_1}, \ldots \theta_{s_q}$ are in general position..Let $\eta_\lambda : N \times V_\lambda \longrightarrow L_\lambda$ be simplifications. Take $a_\lambda \in \mathbb{P}(V_\lambda)$ for $\lambda = 1, \ldots, q$. Then $a_1, \ldots, a_q$ are in general position if and only if $0 \neq \mathfrak{a}_\lambda \in V$ with $\mathbb{P}(\mathfrak{a}_\lambda) = a_\lambda$ and if $\eta_{\mathfrak{a}_1}, \ldots, \eta_{\mathfrak{a}_q}$ are in general position. This definition does not depend on the choice of $\mathfrak{a}_\lambda$. If $a_1, \ldots, a_q$ are in general position, then each η_λ is a semi-amplification and L_λ is adapted to a_λ.

<u>Lemma 16.1.</u> Let L be a holomorphic line bundle on the pure n-dimensional complex manifold N. Let $\eta : N \times V \longrightarrow L$ be a simplification Let $a_1, \ldots, a_q$ in $\mathbb{P}(V)$ be in <u>general position</u> for L with $q > n$. Then η is an amplification.

<u>Proof.</u> Assume $x \in E_L[\infty]$. Then $x \in E_L[a_\lambda]$ for $\lambda = 1, \ldots, q$. Hence $q \leqq n$. Contradiction! Therefore $E_L[\infty] = \emptyset$;q.e.d.

More properties on general position are given in the Appendix.

b) The singular volume form

The following General Assumptions shall be made.

(C1) Let N be a compact, connected, complex manifold of dimension $n > 0$.

(C2) Let $L_1, \dots, L_q$ be holomorphic line bundles on N and let κ_j be a hermitian metric along the fibers of L_j for $j = 1, \dots, q$.

(C3) Let V_j be a linear subspace of $\Gamma(N, L_j)$, let ℓ_j be a hermitian metric on V_j and let $\eta_j \colon N \times V_j \longrightarrow L_j$ be the evaluation map for $j = 1, \dots, q$. Assume $\dim V_j > 0$ and that κ_j is distinguished.

(C4) Given are $a_j \in \mathbb{P}(V_j)$ for $j = 1, \dots, q$ such that $a_1, \dots, a_q$ are in general position. Define $u_j = \|\square, a_j\|^2_{\kappa_j}$ for $j = 1, \dots, q$.

(C5) Define $L = L_1 \otimes \dots \otimes L_q$ and $\kappa = \kappa_1 \otimes \dots \otimes \kappa_q$ as a hermitian metric along the fibers of L. Define $V = V_1 \otimes \dots \otimes V_q \subseteq \Gamma(N, L)$. Then $\ell = \ell_1 \otimes \dots \otimes \ell_q$ is a hermitian metric on V. Define $a = a_1 \otimes \dots \otimes a_q \in \mathbb{P}(V)$. Let $\eta \colon V \longrightarrow L$ be the evaluation map.

Remarks. The function $u_j \geqq 0$ is of class C^∞ with $u_j^{-1}(0) = E_L[a_j]$. Also

(16.2)
$$E_L[a] = E_L[a_1] \cup \dots \cup E_L[a_q].$$

Each η_j is a semi-amplification and L_j is adapted to a_j. Also $\eta = \eta_1 \otimes \dots \otimes \eta_q$ is a semi-amplification and L is adapted to a. Obviously

(16.3)
$$c(L, \kappa) = c(L_1, \kappa_1) + \dots + c(L_q, \kappa_q)$$

(16.4)
$$c(L_j, \kappa_j) = -\, dd^c \log u_j \quad \text{on } N - E_L[a_j]$$

since $\Phi_{L_j}[a_j] = 0$. (Lemma 4.7). Define

(16.3)
$$u = (u_1, \dots, u_q) \colon N \longrightarrow \mathbb{R}^q_+.$$

For $\lambda > 0$ define $P_\lambda : \mathbb{R}(0,1/e)^q \longrightarrow \mathbb{R}_+$ by

$$(16.4) \qquad P_\lambda(\mathfrak{r}) = \prod_{j=1}^{q} x_j^\lambda (\log x_j)^2 .$$

If $\delta > 0$ such that $e\, u_j \delta < 1$, then $P_\lambda(\delta u) = P_\lambda \circ (\delta u) > 0$ is defined and of class C^∞ on $N - E_L[a]$. Write also $P = P_1$. Let K_N be the canonical bundle of N.

Lemma 16.2. Assume (C1) - (C5). Then $L \otimes K_N$ is positive (respectively non-negative) if and only if a form $\Omega > 0$ of degree 2n and class C^∞ exists such that $c(L,\kappa) + \mathrm{Ric}\,\Omega > 0$ (respectively $c(L,\kappa) + \mathrm{Ric}\,\Omega \geqq 0$) on N. If $\lambda > 1$, if $c(L,\kappa) > 0$ on N, and if $L \otimes K_N$ is non-negative, then a form $\Omega > 0$ of degree 2n and class C^∞ exists on N such that $\lambda c(L,\kappa) + \mathrm{Ric}\,\Omega > 0$. If $c(L,\kappa) > 0$ on N, if $L \otimes K_n$ is positive, then $\lambda \in \mathbb{R}(0,1)$ exists such that $\lambda\, c(L,\kappa) + \mathrm{Ric}\,\Omega > 0$.

The proof is easily obtained, see [82] Lemma 19.4. Now an additional General Assumption is made:

(C6) Let $\Omega > 0$ be a form of class C^∞ and degree 2n on N. Let $\lambda > 0$ be given and assume $\lambda\, c(L,\kappa) + \mathrm{Ric}\,\Omega > 0$.

Assume (C1) - (C6). A form ξ_a on $N - E_L[a]$ is called a Carlson-Griffiths form if there exist constants $\delta > 0$ and $\gamma > 0$ such that

$$(16.5) \qquad 0 \leqq \delta\, u_j e < 1 \qquad \text{on } N \text{ for } j = 1,\dots,q$$

$$(16.6) \qquad \xi_a = \frac{\gamma\Omega}{P_\lambda(\delta u)} > 0 \qquad \text{on } N - E_L[a]$$

$$(16.7) \qquad \mathrm{Ric}\,\xi_a > 0 \qquad \text{on } N - E_L[a]$$

$$(16.8) \qquad (\mathrm{Ric}\,\xi_a)^n \geqq \prod_{j=1}^{q} (\delta uj)^{\lambda-1} \xi_a > 0 \quad \text{on } N - E_L[a]$$

$$(16.9) \qquad 0 \leqq \int_N (\mathrm{Ric}\,\xi_a)^n < \infty .$$

Remarks: The multiplication of u_j by δ can be regarded as an adjustment of the metric κ_j not changing $c(L_j,\kappa_j)$. For the Second Main

Theorem, the Third Main Theorem and the defect relation $\lambda = 1$ suffices, but for an application concerning the logarithmic derivative $\lambda > 1$ is handy.

Theorem 16.3. Assume (C1) - (C6). Then a Carlson-Griffiths form exists.

Proof. Since N is compact, and $\chi = \lambda\, c(L,\kappa) + \text{Ric}\,\Omega > 0$, a constant $\gamma_0 > 1$ exists such that

$$(16.10) \qquad - \gamma_0\chi < c(L_j,\kappa_j) < \gamma_0\chi$$

for $j = 1,\ldots,q$ on N. Define $m_j = \text{Max}\,\{u_j(x) \mid x \in N\}$ and $m = \text{Max}(m_1,\ldots,m_q)$. Take $\delta > 0$ so small that

$$(16.11) \qquad 1 < 4\, q\gamma_0 < \log \frac{1}{\delta m}\,.$$

Then $0 < \delta u_j e^{4q\gamma_0} \leqq \delta m(e^{4q\gamma_0}) < 1$ and (16.5) is proved. On $N - E_L[a]$ define $\xi_a^0 = \Omega/P_\lambda(\delta u) > 0$. Then

$$\begin{aligned} \text{Ric}\,\xi_a^0 &= \text{Ric}\,\Omega - dd^c\log P_\lambda(\delta u) \\ &= \text{Ric}\,\Omega + \lambda \sum_{j=1}^{q} c(L_j,\kappa_j) - 2\sum_{j=1}^{q} \frac{c(L_j,\kappa_j)}{|\log \delta u_j|} \end{aligned}$$

$$(16.12) \qquad + 2\sum_{j=1}^{q} \frac{d \log u_j \wedge d^c\log u_j}{(\log \delta u_j)^2}$$

$$\geqq \chi(1-2\gamma_0 \sum_{j=1}^{q} |\log \delta u_j|^{-1}) + 2\sum_{j=1}^{q} \frac{d \log u_j \wedge d^c\log u_j}{(\log \delta u_j)^2}$$

$$\geqq \frac{\chi}{2} + 2\sum_{j=1}^{q} \frac{d \log u_j \wedge d^c\log u_j}{(\log \delta u_j)^2} > 0$$

on $N - E_L[a]$.

Take $x \in N$. The existence of an open neighborhood $U(x)$ of x and a constant $\gamma(x) > 0$ is claimed such that

$$(16.13)\qquad (\mathrm{Ric}\ \xi_a^0)^n \geqq \gamma(x)\Big(\prod_{j=1}^{q}\delta u_j\Big)^{\lambda-1}\xi_a^0 > 0$$

on $U(x) - E_L[a]$. If $x \in N - E_L[a]$, this is trivial. Take $x \in E_L[a]$. Then $p \in \mathbb{N}[1,q]$ and $\mu \in \mathfrak{T}(p,q)$ exist such that $x \in E_{L_j}[a_j]$ if and only if $j \in \mathrm{Im}\,\mu$. Because $a_1,\ldots,a_q$ are in general position $p \leqq n$. Also a smooth patch $\alpha\colon U_\alpha \longrightarrow U'_\alpha$ with $x \in U_\alpha$ and $\alpha(x) = 0$ exists such that $\alpha = (\alpha_1,\ldots,\alpha_n)$ and

$$U_\alpha \cap E_L[a] = U_\alpha \cap \bigcup_{\nu=1}^{p} E_{L_{\mu(\nu)}}[a_{\mu(\nu)}]$$

$$U_\alpha \cap E_{L_{\mu(\nu)}}[a_{\mu(\nu)}] = \alpha_\nu^{-1}(0)$$

for $\nu = 1,\ldots,p$. Also U_α is taken so small that $L_j|U_\alpha$ is trivial for $j = 1,\ldots,q$. Take $s_j \in V_j$ with $|s_j|_{\ell_j} = 1$ and $\mathbb{P}(s_j) = a_j$. Then $\theta_{\alpha_\nu} = \theta_{s_{\mu(\nu)}}$. Hence $s_{\mu(\nu)} = \alpha_\nu v_\nu$ where v_ν is a holomorphic frame of $L_{\mu(\nu)}$ over U_α. Define $b_\nu = |v_\nu|^2_{\kappa_{\mu(\nu)}} > 0$. Then

$$u_{\mu(\nu)} = \|\square, a_{\mu(\nu)}\|^2_{\kappa_{\mu(\nu)}} = |s|^2_{\kappa_{\mu(\nu)}} = |\alpha_\nu|^2 b_\nu$$

on U_α. A form ρ_ν of class C^∞ and bidegree $(1,1)$ on U_α is defined by

$$(16.14)\quad \rho_\nu = \frac{\bar{\alpha}_\nu}{b_\nu}\, d\alpha_\nu \wedge \bar{\partial} b_\nu + \frac{\alpha_\nu}{b_\nu}\, \partial b_\nu \wedge d\bar{\alpha}_\nu + \Big(\frac{|\alpha_\nu|}{b_\nu}\Big)^2 \partial b_\nu \wedge \bar{\partial} b_\nu .$$

If $\alpha_\nu(z) = 0$, then $\rho_\nu(z) = 0$. A form $S_\nu \geqq 0$ of class C^∞ and of bidegree $(1,1)$ is defined on U_α by

$$S_\nu = |\alpha_\nu|^2\, d\log u_{\mu(\nu)} \wedge d^c \log u_{\mu(\nu)}$$

$$= \frac{i}{2\pi}(d\alpha_\nu \wedge d\bar{\alpha}_\nu + \rho_\nu).$$

Observe $S_\nu \wedge S_\nu = 0$. Also (16.12) implies

$$(\mathrm{Ric}\ \xi_a^0)^n \geqq [\frac{\chi}{2} + \sum_{\nu=1}^{p} 2(|\alpha_\nu| \log (\delta u_{\mu(\nu)}))^{-2} S_\nu]^n$$

$$\geqq \frac{n!}{(n-p)!} (\frac{\chi}{2})^{n-p} \wedge 2^p \prod_{\nu=1}^{p} (|\alpha_\nu| \log (\delta u_{\mu(\nu)}))^{-2} S_\nu$$

on $U_\alpha - E_L[a]$. A form ρ of class C^∞ and bidegree (p,p) exists on U_α such that $\rho(x) = 0$ and such that

$$S = \prod_{\nu=1}^{p} S_\nu = (\frac{i}{2\pi})^p d\alpha_1 \wedge d\bar{\alpha}_1 \wedge \dots \wedge d\alpha_p \wedge d\bar{\alpha}_p + \rho.$$

A number $r \in \mathbb{R}(0,1)$ exists such that

$$\Delta = \{(z_1,\dots,z_n) \in \mathbb{C}^n \mid |z_\nu| \leqq r\} \subseteq U'_\alpha.$$

An open neighborhood $U(x)$ of x and a constant $\gamma_1 > 0$ exist such that $\overline{U(x)}$ is compact and contained in U_α and such that

$$\frac{\chi}{2} > \gamma_1 \frac{i}{2\pi} \sum_{\nu=p+1}^{n} d\alpha_\nu \wedge d\bar{\alpha}_\nu$$

$$|\alpha_\nu| < r < 1 \qquad \text{for all } \nu \in \mathbb{N}[1,n]$$

$$|\rho \wedge (\frac{\chi}{2})^{n-p}| < \frac{1}{2}(n-p)!\ \gamma_1^{n-p} (\frac{i}{2\pi})^n \prod_{\nu=1}^{n} d\alpha_\nu \wedge d\bar{\alpha}_\nu$$

on $U(x)$. Then

$$(\frac{\chi}{2})^{n-p} \wedge S \geqq (n-p)!\ \gamma_1^{n-p} (\frac{i}{2\pi})^n \prod_{\nu=1}^{n} d\alpha_\nu \wedge d\bar{\alpha}_\nu + \rho \wedge (\frac{\chi}{2})^{n-p}$$

$$\geqq \frac{1}{2} (n-p)!\ \gamma_1^{n-p} (\frac{i}{2\pi})^n \prod_{\nu=1}^{n} d\alpha_\nu \wedge d\bar{\alpha}_\nu.$$

Since $p \geqq 1$, this implies

$$(\mathrm{Ric}\ \xi_a^0)^n \geqq \gamma_1^{n-p} \prod_{\nu=1}^{p} (\alpha_\nu| \log (\delta u_{\mu(\nu)}))^{-2} (\tfrac{i}{2\pi})^n \prod_{\nu=1}^{n} d\alpha_\nu \wedge d\bar{\alpha}_\nu$$

on $U(x) - E_L[a]$. A constant $\gamma_2 > 0$ exists such that

$$\gamma_1^{n-p} (\tfrac{i}{2\pi})^n \prod_{\nu=1}^{n} d\alpha_\nu \wedge d\bar{\alpha}_\nu > \gamma_2 \Omega$$

on $U(x)$. Constants $\gamma_3 > 0$ and $\gamma_4 > 0$ exist such that

$$\gamma_2 \delta^p \prod_{\nu=1}^{p} b_\nu > \gamma_3 \qquad \text{and } \delta u_j \log (\delta u_j)^2 > \gamma_4$$

for all $j \notin \mathrm{Im}\ \mu$.Therefore

$$(\mathrm{Ric}\ \zeta_a^0)^n \geqq \gamma_2 \prod_{\nu=1}^{p} (|\alpha_\nu| \log \delta u_{\mu(\nu)})^{-2} \Omega$$

$$= \frac{\gamma_2\ \delta^p \prod_{\nu=1}^{p} b_\nu}{\prod_{\nu=1}^{p} \delta u_{\mu(\nu)} (\log \delta u_{\mu(\nu)})^2} \Omega$$

$$\geqq \gamma_3 \gamma_4^{n-p} \frac{\Omega}{\prod_{j=1}^{q} (\delta u_j) \log (\delta u_j)^2}$$

$$= \gamma_3 \gamma_4^{n-p} \prod_{j=1}^{q} (\delta u_j)^{\lambda-1}\ \xi_a^0 > 0$$

on $U(x) - E_L[a]$. Hence (16.13) is satisfied with $\gamma(x) = \gamma_3\gamma_4^{n-p}$.

By Heine-Borel, $N = U(x_1) \cup \ldots \cup U(x_k)$. Define $\gamma = \operatorname*{Min}_{1 \leqq j \leqq k} \gamma(x_j)$.

Then $\gamma > 0$ is constant and $\xi_a = \gamma\ \xi_a^0$ satisfies (16.6),(16.7) and (16.8), since $\mathrm{Ric}\ \xi_a = \mathrm{Ric}\ \xi_a^0$.

Since N is compact, it suffices to show that $(\mathrm{Ric}\ \xi_a)^n$ is integrable over each neighborhood $U(x)$ constructed. If $x \in N - E_L[a]$, this is trivial. Take $x \in E_L[a]$. Define $\gamma_5 = 1 + 2q\gamma_0$. Then (16.5), (16.10) and (16.12) imply

$$\mathrm{Ric}\ \zeta_a \leqq \gamma_5\chi + 2 \sum_{j=1}^{q} (\log \delta u_j)^{-2} d \log u_j \wedge d^c \log u_j .$$

Let Σ^* be the summation over all $(h_1,\ldots,h_q)$ such that $h_j \in \{0,1\}$ for $j = 1,\ldots q$ and such that $0 \leqq h_1+\ldots+h_q = h \leqq n$. Then

$$(\mathrm{Ric}\zeta_a^0)^n \leqq \Sigma^* \frac{n!}{(n-h)!} (\gamma_5\chi)^{n-h} \prod_{j=1}^{q} \left(\frac{2\ d \log u_j \wedge d^c \log u_j}{(\log \delta u_j)^2}\right)^{h_j}.$$

Recall that $\mu\colon \mathbb{N}[1,p] \longrightarrow \mathbb{N}[1,q]$ is an injective, increasing map. Define $A = \mathbb{N}[1,q] - \mathrm{Im}\mu$. If $j \in \mathrm{Im}\mu$, then $j = \mu(\nu)$ for one $\nu \in \mathbb{N}[1,p]$. Hence

$$\log \frac{1}{\delta} + \log \frac{1}{b_\nu} + \log \frac{1}{|\alpha_\nu|^2} = \log \frac{1}{\delta u_j} \geqq 1$$

on U_α. A constant $\gamma_6 > 1$ exists such that $\left| \log \frac{1}{\delta} + \log \frac{1}{b_\nu} \right| < \gamma_6$ on $U(x)$. If $\log |\alpha_\nu| \leqq - \gamma_6$, then

$$\log \frac{1}{\delta u_j} \geqq \log \frac{1}{|\alpha_\nu|^2} - \gamma_6 \geqq \log \frac{1}{|\alpha_\nu|} \geqq \frac{1}{\gamma_6} \log \frac{1}{|\alpha_\nu|} > 0.$$

If $\log |\alpha_\nu| > - \gamma_6$, then $\log |\alpha_\nu| \leqq \log r < 0$ on $U(x)$ and

$$\log \frac{1}{\delta u_j} \geqq 1 \geqq \frac{1}{\gamma_6} \log \frac{1}{|\alpha_\nu|} > 0.$$

Therefore

$$0 \leqq \frac{(\log |\alpha_\nu|^2)^2}{(\log \delta u_j)^2} \leqq \gamma_6^2$$

on $U(x)$. If $j \in A$, then $u_j(z) \neq 0$ for all $z \in \overline{U(x)}$. Hence

$$\Phi_{h_1 \cdots\cdots h_q} = \prod_{j\in A} \left(\frac{2d \log u_j \wedge d^c \log u_j}{(\log \delta u_j)^2}\right)^{h_j} \times$$

$$\prod_{\nu=1}^{p} \left[\frac{2S_\nu (\log |\alpha_\nu|^2)^2}{(\log \delta u_{\mu(\nu)})^2}\right]^{h_{\mu(\nu)}} \times$$

$$\prod_{\nu=1}^{p} \left[|\alpha_\nu|^2 (\log |\alpha_\nu|^2)\right]^{1-h_{\mu(\nu)}}$$

$$\Phi = \Sigma^* \frac{n!}{(n-p)!} (\gamma_5 \chi)^{n-h} \Phi_{h_1 \cdots h_q}$$

have bounded coefficients on U(x) with

$$\Phi \geqq \prod_{\nu=1}^{p} |\alpha_\nu|^2 (\log |\alpha_\nu|^2)^2 (\mathrm{Ric}\ \xi_a)^n .$$

A constant $\gamma_7 > 0$ exists such that

$$0 \leqq (\mathrm{Ric}\ \xi_a)^n \leqq \gamma_7 \prod_{\mu=1}^{p} (|\alpha_\mu| \log |\alpha_\mu|^2)^{-2} \left(\frac{i}{2}\right)^n \prod_{\nu=1}^{n} d\alpha_\nu \wedge d\bar{\alpha}_\nu .$$

Since $\alpha(U(x)) \subseteq \Delta$, this implies

$$\int_{U(x)} (\mathrm{Ric}\ \xi_a)^n \leqq \gamma_7 (\pi\, r^2)^{n-p} \left(\frac{\pi}{2}\right)^p |\log r|^{-p} < \infty .$$

q.e.d.

Compare these results with Griffiths-King [33]. Proposition 6.9 and 6.10, and Carlson and Griffiths [8]. Proposition 2.1. The construction of ξ_a in [8] was of the fundamental importance.

17. The Second Main Theorem

The following additional <u>General Assumptions</u> shall be made:

(C7) Let ξ_a be a Carlson-Griffiths form for the λ of (C6).

(C8) Let τ be a parabolic exhaustion of the connected complex manifold M of dimension m.

(C9) Let $f: M \longrightarrow N$ be a holomorphic map.

(C10) Assume that f is adapted to a_j for L_j for $j = 1,\dots,q$.

(C11) Let F be an effective Jacobian section of f.

Assume (C1) - (C11). On $Z_a = M^+ - f^{-1}(E_L[a])$ define $\zeta_a: Z_a \longrightarrow \mathbb{R}_+$ by $F[\xi_a] = \zeta_a \upsilon^m$.

<u>Theorem 17.1.</u> <u>Second Main Theorem.</u> Assume (C1) - (C11). Then $\zeta_a > 0$ on $Z_a - Z(F)$ and

(17.1)
$$\mu_a(r) = \tfrac{1}{2}\int_{M\langle r\rangle} \log \zeta_a \, \sigma$$

exists for all $r \in \mathfrak{C}^0_\tau$. If $0 < s < r \in \mathbb{R}$, the Ricci function Ric (r,s,ξ_a,f) exists. If $0 < s < r \in \mathfrak{C}^0_\tau$ and $s \in \mathfrak{C}^0_\tau$, then

(17.2)
$$\mu_a(r) - \mu_a(s) + \mathrm{Ric}_\tau(r,s) + \lambda \sum_{j=1}^{q} N_{a_j}(r,s,L_j,f)$$
$$= N_{\theta_F}(r,s) + \mathrm{Ric}\,(r,s,\xi_a,f).$$

<u>Proof.</u> Define $v: M - Z(F) \longrightarrow \mathbb{R}_+$ by $\upsilon^m = vF(\Omega)$. Then $F[\xi_a] = \zeta_a vF[\Omega]$. Trivially $\zeta_a > 0$ on $Z_a - Z(F)$. Also

$$F[\xi_a] = \gamma \prod_{j=1}^{q} (\delta u_j \circ f)^\lambda (\log \delta u_j \circ f)^2)^{-1} F[\Omega].$$

As in Appendix (7) define $\mathfrak{L}(x) = \log(\log x)^2$ for $x > 0$. Then

$$\log \zeta_a = -\log v - \lambda \sum_{j=1}^{q} \log \|f,a_j\|^2_{\kappa_j}.$$
$$- \sum_{j=1}^{q} \mathfrak{L}(\delta u_j \circ f) + \log \gamma - q\lambda \log \delta.$$

If $r \in \mathfrak{E}^0_\tau$, each term is integrable over $M\langle r\rangle$ and

$$\mu_a(r) = -\tfrac{1}{2}\int_{M\langle r\rangle} \log v\sigma + \lambda\sum_{j=1}^{q} m_{a_j}(r,L_j,\kappa_j,f)$$

(17.3)

$$-\tfrac{1}{2}\sum_{j=1}^{q}\int_{M\langle r\rangle} \mathfrak{L}(\delta u_j\circ f)\,\sigma + \log\gamma - q\lambda\log\delta.$$

Here Theorem A13 inthe Appendix was used. Take $\mathfrak{a}_j \in V_j$ with $|\mathfrak{a}_j| = 1$ and with $\mathbb{P}(\mathfrak{a}_j) = a_j$. Define $s_j = \eta_{\mathfrak{a}_j}$. Then $u_j = |s_j|^2_{\kappa_j}$. Hence $\delta u_j\circ f = |s_j\circ f|^2_{\delta\kappa_j}$ and Theorem 13.A applies.

(17.4)

$$\int_{M\langle r\rangle} \mathfrak{L}(\delta u_j\circ f)\,\sigma - \int_{M\langle s\rangle} \mathfrak{L}(\delta u_j\circ f)\,\sigma =$$

$$= 2\int_s^r \int_{M[t]} dd^c\mathfrak{L}(\delta u_j\circ f)\wedge \upsilon^{m-1}\, t^{1-2m}dt.$$

If $0 < s < r \in \mathfrak{E}^0_\tau$ and $s \in \mathfrak{E}^0_\tau$, then (12.28), Theorem 15.5 and (17.3) and (17.4) imply

$$\mu_a(r) - \mu_a(s) + \mathrm{Ric}_\tau(r,s) + \lambda\sum_{j=1}^{q} N_{a_j}(r,s,L_j,f_j)$$

(17.5)

$$= N_{\theta_F}(r,s) + \mathrm{Ric}(r,s,\Omega,f) + \lambda\sum_{j=1}^{q} T(r,s,L_j,\kappa_j,f)$$

$$-\sum_{j=1}^{q}\int_s^r\int_{M[t]} dd^c\mathfrak{L}(\delta u_j\circ f)\wedge\upsilon^{m-1}\,t^{1-2m}\,dt.$$

Also

$$f^*(\mathrm{Ric}\,\xi_a) = \mathrm{Ric}\,F[\xi_a] = \mathrm{Ric}\,F[\Omega] - dd^cP_\lambda(\delta u_j\circ f)$$

$$(17.6) \quad = f^*(\mathrm{Ric}\,\Omega) - \lambda \sum_{j=1}^{q} dd^c \log u_j \circ f - \sum_{j=1}^{q} dd^c \mathfrak{L}(\delta u_j \circ f)$$

$$= f^*(\mathrm{Ric}\,\Omega) + \lambda \sum_{j=1}^{q} f^*(c(L_j, \kappa_j)) - \sum_{j=1}^{q} dd^c \mathfrak{L}(\delta u_j \circ f).$$

Therefore

$$\mathrm{Ric}\,(r,s,\xi_a,f) =$$

$$(17.7) \quad = \mathrm{Ric}\,(r,s,\Omega,f) + \lambda \sum_{j=1}^{q} T(r,s,L_j,\kappa_j,f)$$

$$- \sum_{j=1}^{q} \int_s^r \int_{M[t]} dd^c \mathfrak{L}(\delta u_j \circ f) \upsilon^{m-1} t^{1-2m}\, dt.$$

Now (17.5) and (17.7) prove (17.2); q.e.d.

The Second Main Theorem 17.1 interprets and extends Proposition 6.23 of Griffiths-King [33]. Observe they assume $m \geqq n$ which is not needed here. In the associated mapping approach (see [80] and Wong [90]) Theorem 17.1 corresponds to the Plückert Difference Formula. Since ξ_a is singular on $E_L[a]$, the following assymptotic results have still to be proved.

Proposition 17.2. Assume (C1) - (C11). Then $\mathrm{ric}(t,\xi_a,f) \geqq 0$ (see (15.7) increases and is semicontinuous from the right. The limits

$$(17.8) \quad 0 \leqq \mathrm{ric}(0,\xi_a,f) = \lim_{0<t\to 0} \mathrm{ric}(t,\xi_a,f)$$

$$(17.9) \quad 0 \leqq \mathrm{ric}(\infty,\xi_a,f) = \lim_{t\to\infty} \mathrm{ric}(\infty,\xi_a,f)$$

exist with

$$(17.10) \quad \mathrm{ric}(t,\xi_a,f) = \int_{M[t]} f^*(\mathrm{Ric}\ \xi_a) \wedge \omega^{m-1} + \mathrm{ric}(0,\xi_a,f)$$

$$(17.11) \quad \lim_{r\to\infty} \frac{\mathrm{Ric}(r,s,\xi_a,f)}{\log r} = \mathrm{Ric}\,(\infty,\xi_a,f) \leqq +\infty.$$

Proof. Trivally $\mathrm{ric}(t,\xi_a,f) \geqq 0$ is semi-continuous from the right. Define

$$\psi = f^*(\mathrm{Ric}\ \Omega) + \lambda \sum_{j=1}^{q} f^*(c(L_j,\kappa_j)).$$

Take $0 < s < r \in \mathfrak{E}_\tau$ with $s \in \mathfrak{E}_\tau$. Observe $d\psi = 0$. Then

$$\int_{M(s,r]} \psi \wedge \omega^{m-1} = \int_{M(s,r]} d(\psi \wedge d^c \log \tau \wedge \omega^{m-1})$$

$$= \int_{M\langle r\rangle} \psi \wedge d^c \log \tau \wedge \omega^{m-1} - \int_{M\langle s\rangle} \psi \wedge d^c \log \tau \wedge \omega^{m-1}$$

$$= r^{2-2m} \int_{M\langle r\rangle} \psi \wedge d^c\tau \wedge \upsilon^{m-1} - s^{2-2m} \int_{M\langle s\rangle} \psi \wedge d^c\tau \wedge \upsilon^{m-1}$$

$$= r^{2-2m} \int_{M[r]} \psi\, \upsilon^{m-1} - s^{2-2m} \int_{M[s]} \psi\, \upsilon^{m-1}.$$

Semi-continuity implies

$$(17.12)\quad \int_{M(s,r]} \psi \wedge \omega^{m-1} = r^{2-2m} \int_{M[r]} \psi \wedge \upsilon^{m-1} - s^{2-2m} \int_{M[s]} \psi \wedge \upsilon^{m-1}$$

for all $0 < s < r \in \mathbb{R}^+$. Also Appendix (36) with $v = \delta u_j \circ f$ applies and yields with (17.7) and (15.6) the identity:

$$\mathrm{ric}(r,\xi_a,f) - \mathrm{ric}(s,\xi_a,f) =$$

$$= r^{2-2m} \int_{M[r]} f^*(\mathrm{Ric}\ \xi_a) \wedge \upsilon^{m-1} - s^{2-2m} \int_{M[s]} f^*(\mathrm{Ric}\ \xi_a) \wedge \upsilon^{m-1}$$

$$= \int_{M(s,r]} \psi \wedge \omega^{m-1} + \int_{M(s,r]} f^*(dd^c \mathfrak{L}(\delta u_j \circ f)) \wedge \omega^{m-1}$$

$$= \int_{M(s,r]} f^*(\xi_a) \wedge \omega^{m-1} \geqq 0$$

if $0 < s < r$. Hence $\mathrm{ric}(r,\xi_a,f)$ increases and (17.8) - (17.11) follow easily; q.e.d.

Compare the results of this section with Griffiths-King [33] 6(b) and Carlson-Griffiths [8] 3.

VI. The Defect Relation

18. The Third Main Theorem

a) Jacobian sections dominated by the exhaustion

Let M be a complex manifold of pure dimension $m > 0$. Let $\psi > 0$ and χ be forms of class C^k and degree 2m on an open subset $U \neq \emptyset$ of M. Then a function H of class C^k exists uniquely such that $\chi = H\,\psi$ on U. Write $H = \chi/\psi$.

In this section, assume that M and N are complex manifolds of pure dimensions m and n respectively, that τ is a parabolic exhausion of M and that $f: M \longrightarrow N$ is a holomorphic map. A Jacobian section F of f is said to be dominated by τ, if and only if for every $r > 0$ there exists a constant $C \geqq 1$ such that the following property holds:

(P) Let U be open in N. Define $\tilde{U} = f^{-1}(U)$. Assume $\tilde{U} \cap M^+(r) \neq \emptyset$

Then

$$n \left(\frac{F[\psi^n]}{\upsilon^m}\right)^{\frac{1}{n}} \upsilon^m \leqq C\, f^*(\psi) \wedge \upsilon^{m-1} \tag{18.1}$$

holds on $\tilde{U} \cap M^+(r)$ for all continuous forms $\psi \geqq 0$ of bidegree (1,1) on U.

Recall $M^+ = \{x \in M \mid \upsilon^m(x) > 0\}$. Hence $M^+(r) \neq \emptyset$ for all $r > 0$, since τ is parabolic. Let $Y(r)$ be the infimum of all those constants $C \geqq 1$ such that (18.1) holds. Then $Y(r) \geqq 1$ with

$$n \left(\frac{F[\psi^n]}{\upsilon^m}\right)^{\frac{1}{n}} \upsilon^m \geqq Y(r)\, f^*(\psi) \wedge \upsilon^{m-1} \tag{18.2}$$

on $\tilde{U} \cap M^+(r)$. The function $Y = Y_F$ is called the dominator of F. The function Y increases.

Lemma 18.1. Let F be an effective Jacobian section of f dominated by τ. Assume M is connected. Then $m \geqq n = \operatorname{rank} f$.

Proof. Recall rank f = Max $\{\text{rank}_x f \mid x \in M\}$ in the sense of Remmert. If $\rho(x)$ is the rank of the Jacobian of f at x, Remmert [55] showed rank f = Max $\{\rho(x) \mid x \in M\}$. Define p = rank f. Then $p \leqq \text{Min}(m,n)$. Assume $p < n$. The set $E = \{x \in M \mid \rho(x) < p\}$ is thin analytic in M. Also Z(F) is thin. Take $x \in M^+ - (E \cup Z(F))$. Take $r > 0$ such that $x \in M(r)$. Since $\rho(x) = p$, smooth patches $\alpha: U_\alpha \longrightarrow U'_\alpha$ with $x \in U_\alpha$ and $\beta: U_\beta \longrightarrow U'_\beta$ with $f(U_\alpha) \subseteq U_\beta$ exist such that $f^*(d\beta_\mu) = d\alpha_\mu$ for $\mu = 1,\ldots,p$ and $f^*(d\beta_\mu) = 0$ for $\mu = p+1,\ldots n$. Then $F[d\beta] = F_{\beta\alpha} d\alpha$ (see section 14a) where $F_{\beta\alpha}$ is holomorphic on U_α with $F_{\beta\alpha}(x) \neq 0$. Functions $g_{\mu\nu}$ of class C^∞ exist on U_α such that

$$\upsilon = \frac{i}{2\pi} \sum_{\mu,\nu=1}^{m} g_{\mu\nu}\, d\alpha_\mu \wedge d\bar{\alpha}_\nu \qquad \upsilon^m = i_m\, g\, d\alpha \wedge d\bar{\alpha}$$

where $g = \det g_{\mu\nu} > 0$ on $U_\alpha \cap M^+$. Also

$$\upsilon^{m-1} = \left(\frac{i}{2\pi}\right)^{m-1} m! \sum_{\mu\neq\nu} h_{\mu\nu}\, d\alpha_\nu \wedge d\bar{\alpha}_\mu \wedge \prod_{\mu\neq\lambda\neq\nu} d\alpha_\lambda \wedge d\bar{\alpha}_\lambda$$

$$+ \left(\frac{i}{2\pi}\right)^{m-1} m! \sum_{\mu=1}^{m} h_{\mu\mu} \prod_{\mu\neq\lambda} d\alpha_\lambda \wedge d\bar{\alpha}_\lambda$$

with $h_{\mu\mu} > 0$ on $U_\alpha \cap M^+$. Because $x \in U_\alpha \cap M^+(r)$, a constant $\lambda > 0$ exists such that

$$(18.3) \qquad \left(Y(r) \sum_{\mu=1}^{p} h_{\mu\mu}(x)\right)^n < n^n\, |F_{\beta\alpha}(x)|^2\, g(x)^{n-1} \lambda^n.$$

Define

$$\psi = \frac{i}{2\pi} \left(\sum_{\nu=1}^{n-1} d\beta_\nu \wedge d\bar{\beta}_\nu + \lambda^n\, d\beta_n \wedge d\bar{\beta}_n \right)$$

Then $\psi^n = i_n \lambda^n\, d\beta \wedge \bar{\beta}$ and $F[\psi^n]\, g = \lambda^n |F_{\beta\alpha}|^2 \upsilon^m$

Hence

$$n\left(\frac{F[\psi^n]}{\upsilon^m}\right)^{\frac{1}{n}}\upsilon^m = n\,\lambda|F_{\beta\alpha}|^{\frac{2}{n}}\,g^{1-\frac{1}{n}}\,i_m d\alpha \wedge d\bar{\alpha}.$$

Now, $p < n$ implies $f^*(\psi) = \frac{i}{2\pi}(d\alpha_1 \wedge d\bar{\alpha}_1 + \ldots + d\alpha_p \wedge d\bar{\alpha}_p)$ and

$$f^*(\psi) \wedge \upsilon^{m-1} = \sum_{\nu=1}^{p} h_{\nu\nu}\, i_m d\alpha \wedge d\bar{\alpha}.$$

Now, (18.2) implies

$$Y(r)\sum_{\nu=1}^{p} h_{\nu\nu}(x) \geqq n\,\lambda|F_{\alpha\beta}(x)|^{\frac{2}{n}}\,g(x)^{1-\frac{1}{n}}$$

which contradicts (18.3). Therefore $n = p = \operatorname{rank} f \leqq m$; q.e.d.

This remark destroys all hope of finding an easy trick to extend the Carlson-Griffiths-King method to the case $m < n$. If $m \geqq n = r = \operatorname{rank} f$, the existence of effective Jacobian sections dominated by τ will be proved, if $\upsilon > 0$ or if the exhaustion τ is obtained by a covering map from $\mathbb{C}^m$.

Lemma 18.2. Let M be a complex manifold of pure dimension $m > 0$. Take $n \in \mathbb{N}[1,m]$. Define $q = m - n \geqq 0$. Let $\psi \geqq 0$ and $\chi > 0$ forms of bidegree (1,1) on M with $\psi^{n+1} = 0$. Then

$$n\left[\binom{m}{q}\frac{\psi^n \wedge \chi^q}{\chi^m}\right]^{\frac{1}{n}}\chi^m \leqq m\psi \wedge \chi^{m-1}.$$

Proof. Take $a \in M$. A smooth patch $\alpha: U_\alpha \longrightarrow U'_\alpha$ exists such that $a \in U_\alpha$ and $\alpha = (\alpha_1,\ldots,\alpha_m)$ and such that

$$\chi(a) = \frac{i}{2\pi}\sum_{\mu=1}^{m} d\alpha_\mu \wedge d\bar{\alpha}_\mu \qquad \psi(a) = \frac{i}{2\pi}\sum_{\mu=1}^{m} e_\mu d\alpha_\mu \wedge d\bar{\alpha}_\mu$$

with $e_1 \geqq e_2 \geqq \ldots \geqq e_m \geqq 0$. Take $p \in \mathbb{N}[n+1,m]$. Then

$$\psi(a) \geqq e_p \frac{i}{2\pi} \sum_{\mu=1}^{p} d\alpha_\mu \wedge d\bar{\alpha}_\mu$$

$$0 = \psi(a)^p \geqq e_p^p \left(\frac{i}{2\pi}\right)^p p!\, d\alpha_1 \wedge d\bar{\alpha}_1 \wedge \ldots \wedge d\alpha_p \wedge d\bar{\alpha}_p .$$

Hence $e_p = 0$ and $\psi = \frac{i}{2\pi}(e_1 d\alpha_1 \wedge d\bar{\alpha}_1 + \ldots + e_n d\alpha_n \wedge d\bar{\alpha}_n)$. Let t be a real variable. Then

$$\frac{(t\chi(a) - \psi(a))^m}{t^q \chi(a)^m} = \prod_{\mu=1}^{n} (t - e_\mu) = \sum_{\mu=0}^{n} (-1)^\mu \binom{m}{\mu} \frac{\psi(a)^\mu \wedge \chi(a)^{m-\mu}}{\chi(a)^m} t^{n-\mu} .$$

Hence

$$m\, \psi \wedge \chi^{m-1} = \left(\sum_{\mu=1}^{n} e_\mu\right) \chi^m \geqq n \left(\prod_{\mu=1}^{n} e_\mu\right)^{\frac{1}{n}} \chi^m$$

$$= n\left(\binom{m}{q} \frac{\psi^n \wedge \chi^q}{\chi^m}\right)^{\frac{1}{n}} \chi^m$$

at a; q.e.d.

Again let (M, τ) be a parabolic manifold of dimension m and let $f: M \longrightarrow N$ be a holomorphic map. Take $q \in \mathbb{Z}[0.m]$. A holomorphic form φ of degree q on M is said to be dominated by τ if for each $r > 0$ a constant $C_r \geqq 1$ exists such that $i_q \varphi \wedge \bar{\varphi} \leqq (C_r/m)^{m-q} \upsilon^q$ on $M(r)$. Let $Y(r)$ be the infimum of all those constants $C_r \geqq 1$ for each fixed r. Then $Y(r) \geqq 1$ with

$$(18.4) \qquad i_q \varphi \wedge \bar{\varphi} \leqq \left(\frac{Y(r)}{m}\right)^{m-q} \upsilon^q \qquad \text{on } M(r).$$

The function $Y = Y_\varphi$ increases and is called the dominator of φ. Obviously, if $\upsilon > 0$ or if $q = 0$, then φ is dominated by τ. If $q = 0$ and $\varphi = 1$, then $Y_\varphi = m$ is constant.

Lemma 18.3. Let φ be a holomorphic form of degree $q = m - n \geqq 0$ on M dominated by τ. Then the induced Jacobian section F_φ is dominated by τ with $Y_{F_\varphi} = Y_\varphi$.

Proof. Let U be open in M. Define $\tilde{U} = f^{-1}(U)$. Take $r > 0$. Assume $\tilde{U} \cap M^+(r) \neq \emptyset$. Take a continuous form $\psi \geqq 0$ of bidegree (1,1) on U. Then (14.6) implies

$$F_\varphi[\psi^n] = \binom{m}{q} i_q \, \varphi \wedge \overline{\varphi} \wedge f^*(\psi^n) \leqq \binom{m}{q} \left(\frac{Y(r)}{m}\right)^n \upsilon^q \wedge f^*(\psi^n).$$

Since $f^*(\psi^{n+1}) = 0$, Lemma 18.2 implies

$$n\left(\frac{F_\varphi[\psi^n]}{\upsilon^m}\right)^{\frac{1}{n}} \upsilon^m \leqq n \, \frac{Y(r)}{m} \left(\binom{m}{q} \frac{\upsilon^q \wedge f^*(\psi^n)}{\upsilon^m}\right)^{\frac{1}{n}} \upsilon^m$$

$$\leqq Y(r) \; f^*(\psi) \wedge \upsilon^{m-1}$$

on $\tilde{U} \cap M^+(r)$. Hence F_φ is dominated by τ with $Y_{F_\varphi} = Y = Y_\varphi$; q.e.d.

Corollary 18.4. If $m = n$ and if f has strict rank m, then the Jacobian section F defined by df is effective and dominated by τ with $Y \equiv m$, (Set $\varphi=1$).

Proposition 18.5. If $q = m - n \geqq 0$, if $\upsilon > 0$ on M and if f has strict rank n, there exists a holomorphic form φ of degree q on M which is effective for f and dominated by τ.

Proof. Since $\upsilon > 0$, the manifold M is Stein. Then φ exists by Theorem 14.1 and φ is dominated by τ because $\upsilon > 0$, q.e.d.

Proposition 18.6. Assume $q = m - n \geqq 0$. Assume f has strict rank n. Let $\beta: M \longrightarrow \mathbb{C}^m$ be a holomorphic map of strict rank m. Assume $\tau = |\beta|^2$. Then there exists a holomorphic form φ of degree q on M which is effective for f and dominated by τ with a constant dominator $Y \equiv m$. Moreover, if M is connected, then $\mu \in \mathfrak{T}(q,m)$ exists such that $\varphi = d\beta_\mu$ is such a form.

Proof. W.ℓ.o.g. M can be assumed to be connected. Thin analytic sets S of $\mathbb{C}^m$ and $\tilde{S} = \beta^{-1}(S)$ of M exist such that f has pure Jacobian rank n on $M_0 = M - S$ and such that $\beta: M_0 \longrightarrow (\mathbb{C}^m - S)$ is proper, surjecti and locally biholomorphic.

Take $a \in M_0$. Take a smooth patch $\alpha: U_\alpha \longrightarrow U'_\alpha$ with $f(a) \in U_\alpha$ and an open, connected neighborhood U_β of a in M_0 with $f(U_\beta) \subseteq U_\alpha$ such that $\beta: U_\beta \longrightarrow U'_\beta$ is biholomorphic. Define $f_\nu = \alpha_\nu \circ f$. Then

$$df_\nu = \sum_{\mu=1}^{m} f_{\nu\mu} d\beta_\mu .$$

If $\mu \in \mathfrak{T}(b,c)$ and $\nu \in \mathfrak{T}(k,\ell)$ define $f_{\mu\nu} = \det(f_{\mu(i)\nu(j)})$. Let $\iota: \mathbb{N}[1,n] \to \mathbb{N}[1,n]$ be the identity. Then

$$0 \neq df_1 \wedge \ldots \wedge df_n = \sum_{\mu \in \mathfrak{T}(n,m)} f_{\iota\mu} d\beta_\mu .$$

Hence $\mu \in \mathfrak{T}(n,m)$ exists such that $f_{\iota\mu} \neq 0$ on U_β. Define $\lambda = \mu^\perp \in \mathfrak{T}(q,m)$. Then $\varphi = d\beta_\lambda$ is a holomorphic form of degree q on M. Let F_φ be the induced Jacobian section. Then

$$F_\varphi[d\alpha] = \varphi \wedge f^*(d\alpha) = d\beta_\lambda \wedge df_1 \wedge \ldots \wedge df_n = f_{\iota\mu} d\beta_\lambda \wedge d\beta_\mu$$

$$= f_{\iota\mu} \operatorname{sign}(\lambda,\mu)\, d\beta \not\equiv 0.$$

on U_β. Now $Z(F_\varphi) \cap U_\beta = Z(F_\varphi[d\alpha]) \cap U_\beta = Z(f_{\iota\mu})$ is thin in U_β. Since M is connected, $Z(F_\varphi)$ is thin and F_φ and φ are effective. Observe

$$\upsilon^q = \left(\frac{i}{2\pi}\right)^q q! \sum_{\nu \in \mathfrak{T}(q,m)} d\beta_{\nu(1)} \wedge d\overline{\beta}_{\nu(1)} \wedge \ldots \wedge d\beta_{\nu(q)} \wedge d\overline{\beta}_{\nu(q)}$$

$$\geqq i_q\, d\beta_\lambda \wedge d\overline{\beta}_\lambda = i_q\, \varphi \wedge \overline{\varphi}.$$

Hence φ is dominated by τ with $Y = m$ as the dominator, q.e.d.

b) Estimates of the Ricci function of ξ_a

An additional General Assumption will be made:

(C12) The effective Jacobian section F of f is dominated by τ with dominator Y.

Assume (C1) - (C12) with $\lambda = 1$. Define Z_a, ζ_a and μ_a as in section 17. For $t > 0$ and $0 < s < r$ define

(18.5) $$0 \leqq Q_a(t) = n\, t^{2-2m} \int_{M[t]} (\zeta_a)^{\frac{1}{n}} \upsilon^m \leqq +\infty$$

(18.6) $$0 \leqq S_a(r,s) = \int_s^r Q_a(t) \frac{dt}{t} \leqq +\infty.$$

Lemma 18.7. Assume (C1) - (C12) with $\lambda = 1$. Take $0 < s < r$. Then

(18.7) $$0 \leqq S_a(r,s) \leqq Y(r)\, \mathrm{Ric}(r,s,\xi_a,f) < \infty.$$

Proof. (16.8) with $\lambda = 1$ implies $F[(\mathrm{Ric}\ \xi_a)^n] \geqq F[\xi_a] = \zeta_a \upsilon^m$. Take $r > 0$, then

$$Y(r)\, f^*(\mathrm{Ric}\ \xi_a) \wedge \upsilon^{m-1} \geqq n\left(\frac{F[(\mathrm{Ric}\ \xi_a)^n]}{\upsilon^m}\right)^{\frac{1}{n}} \upsilon^m \geqq n(\zeta_a)^{\frac{1}{n}} \upsilon^m$$

on $M(r) \wedge Z_a$. Hence

$$Y(r)\, \mathrm{Ric}(r,s,\xi_a,f) = \int_s^r t^{1-2m} \int_{M[t]} Y(r)\, f^*(\mathrm{Ric}\ \xi_a) \wedge \upsilon^{m-1}\, dt$$

$$\geqq \int_s^r t^{1-2m} \int_{M[t]} n(\zeta_a)^{\frac{1}{n}} \upsilon^m dt = S_a(r,s)$$

q.e.d.

Lemma 18.8. Assume (C1) - (C12) with $\lambda = 1$. Then $0 < Q_a(t) < \infty$ for all $t > 0$ and $0 < S_a(r,s) < \infty$ if $0 < s < r$. Also $0 < \mathrm{ric}(r,\xi_a,f) < \infty$ if $r > 0$ and

$$(18.8)\qquad 0 < \lim_{r\to\infty} \frac{\mathrm{Ric}(r,s,\xi_a f)}{\log r} = \mathrm{ric}(\infty,\xi_a,f) \leqq +\infty$$

$$(18.9)\qquad \mathrm{Ric}(r,s,\xi_a,f) \longrightarrow \infty \qquad \text{for } r \longrightarrow \infty.$$

Proof. $t^{2m-2}\, Q_a(t)$ increases and $S_a(r,s) < \infty$ by (18.7). Hence $Q_a(t) < +\infty$ for all $t > 0$. If $t>0$, then $M^+(t) \neq \emptyset$ and $\zeta_a > 0$ outside a thin analytic subset of $M^+(t)$. Hence $Q_a(t) > 0$ and $S_a(r,s) > 0$. Since $\mathrm{ric}(r,\xi_a,f)$ increases the remaining assertions follow trivially; q.e.d.

Lemma 18.8. Assume (C1) - (C12) with $\lambda = 1$. Then

$$(18.10)\qquad \hat{Q}_a(t) = \frac{1}{\varsigma} \int_{M\langle t\rangle} (\zeta_a)^{\frac{1}{n}}\, \sigma \geqq 0$$

exists for almost all $t > 0$. If $r > 0$, then

$$(18.11)\qquad r^{2m-2} Q_a(r) = 2mn\,\varsigma \int_0^r \hat{Q}_a(t)\; t^{2m-1}\, dt.$$

For almost all $r > 0$

$$(18.12)\qquad \frac{d}{dr}[r^{2m-2} Q_a(r)] = 2mn\,\varsigma\; \hat{Q}_a(r)\; r^{2m-1}$$

$$(18.13)\qquad 2\mu_a(r) \leqq \varsigma\, n \log \hat{Q}_a(r)$$

(Carlson-Griffiths [8] Lemma 4.5).

Proof. Take $r > 0$. Then

$$r^{2m-2} Q_a(r) = n\, m \int_{M[r]} (\zeta_a)^{\frac{1}{n}}\, \tau^{m-1} d\tau \wedge \sigma =$$

$$= 2n\, m \int_0^r \int_{M\langle t\rangle} (\zeta_a)^{\frac{1}{n}}\, \sigma\; t^{2m-1}\, dt.$$

Hence $\hat{Q}_a$ exist for almost all $t > 0$ and (18.11) holds, which implies

(18.12). Now

$$2\mu_a(r) = \int_{M\langle r\rangle} \log \zeta_a \sigma = \varsigma n \left(\frac{1}{\varsigma} \int_{M\langle r\rangle} \log (\zeta_a)^{\frac{1}{n}} \sigma\right)$$

$$\leqq \varsigma n \log \frac{1}{\varsigma} \int_{M\langle r\rangle} (\zeta_a)^{\frac{1}{n}} \sigma = \varsigma n \log \hat{Q}_a (r).$$

for almost all $r > 0$;q.e.d.

The following Lemma is well-known Nevanlinna [49] 209.

Lemma 18.10. Let $f \geqq 0$, $g \geqq 0$ and $\alpha > 0$ be increasing, continuous functions on $\mathbb{R}^+$. Assume g is of class C^1 on $\mathbb{R}_+$ and

$$\int_s^\infty \frac{dx}{\alpha(x)} < \infty$$

for all $s > 0$. Then a measurable subset E of $\mathbb{R}^+$ exists such that $0 \leqq \int_E g' \, dx < \infty$ and such that $0 \leqq f'(x) \leqq g'(x)\, \alpha(f(x))$ if $x \in \mathbb{R}^+ - E$.

Take $\varepsilon > 0$. A measurable subset E of $\mathbb{R}^+$ is said to be of finite ε-measure if $\int_E x^\varepsilon dx < +\infty$. If f and h are functions on $\mathbb{R}^+$, then $f = h$ or $f \leqq h$ is said to hold except for a set of finite ε-measure if and only if a measurable subset E of $\mathbb{R}^+$ exists such that E has finite ε-measure and such that $f(x) = h(x)$ (respectively $f(x) \leqq h(x)$) for all $x \in \mathbb{R}^+ - E$. If so, write

$$\|_\varepsilon \; f(x) = g(x) \qquad\qquad \|_\varepsilon \; f(x) \leqq h(x).$$

Lemma 18.11. Assume (C1) - (C12) with $\lambda = 1$. Take $\varepsilon > 0$. Then $\nu > 1$ exists such that

$$(18.14) \qquad \|_\varepsilon \; 2\mu_a(r) \leqq 2\nu n \varsigma \log S_a(r,s) + 3\varepsilon\varsigma n \log r$$

$$- \nu n \varsigma \log (2mn\varsigma)$$

$$(18.15)\qquad \|_{\varepsilon}\quad 2\mu_a(r) \leqq (2+\varepsilon)\varsigma\, n \log \mathrm{Ric}\,(r,s,\xi_a,f) + (2+\varepsilon)\varsigma\, n \log Y(r) + 4\varepsilon\varsigma\, n \log r.$$

Proof. Take $\beta \in \mathbb{R}(0,\varepsilon)$ such that $\varepsilon(\beta + 2) - \beta(2m - 1) < 4\varepsilon$ and such that $\beta < 1$. Define $\nu = 1 + \beta/2 > 1$. Then $\varepsilon(\nu + 1) + (\nu - 1)(2m - 1) < 3\varepsilon$ and $\nu < 2$. Lemma 18.10 with $(1 + \varepsilon)\, g(x) = x^{1+\varepsilon}$ and $\alpha(x) = x^{\nu}$ implies

$$\|_{\varepsilon}\qquad \hat{Q}_a(x)\, x^{2m-1} \leqq x^{\varepsilon+2\nu(m-1)} \frac{Q_a(x)^{\nu}}{(2mn\varsigma)^{\nu}}$$

$$\|_{\varepsilon}\qquad Q_a(x) \leqq x^{1+\varepsilon}\, (S_a(x,s))^{\nu}$$

Therefore

$$\|_{\varepsilon}\qquad \hat{Q}_a(x) \leqq x^{\varepsilon+2\nu(m-1)+\nu(1+\varepsilon)-2m+1} \frac{S_a(x,s)^{\nu^2}}{(2mn\varsigma)^{\nu}}$$

with $\varepsilon + 2\nu(m-1) + \nu(1+\varepsilon) - 2m + 1 \leqq \varepsilon(\nu+1) + (\nu-1)(2m-1) < 3\varepsilon$. Because $\mathbb{R}(0.1]$ has finite ε-measure, and because $\nu^2 \leqq 2\nu$ this implies

$$\begin{aligned} \|_{\varepsilon}\qquad 2\mu_a(r) &\leqq \varsigma\, n \log \hat{Q}_a(r) \\ &\leqq 2\varsigma\, n\, \nu \log S_a(r,s) + 3\varsigma\,\varepsilon\, n \log r \\ &\quad - \varsigma\, n\, \nu \log (2mn\varsigma) \end{aligned}$$

which is (18.14). If r large enough, then $-\nu \log 2mn\varsigma \leqq \varepsilon \log r$. Also $2\nu = 2 + \beta < 2 + \varepsilon$ and $\mathrm{Ric}(r,s,\xi_a,f) \geqq 1$ for sufficiently large r. Observe $Y(r) \geqq 1$. Hence (18.7) and (18.14) give (18.15); q.e.d.

The Lemma at 18.7 - 18.11 also hold for $0 < \lambda \leqq 1$, but with little consequence. The following result is valid for all $\lambda > 0$.

Proposition 18.12. Assume (C1) - (C12) with $\lambda > 0$. Define $c_3 = q\varsigma\, |\log 2/q\varsigma|$. Take $s > 0$. Then there are constants $c_1 \leqq 0$ and $c_2 \geqq 0$ such that

(18.16) $$c_1 \leqq \lambda\, T(r,s,L,\kappa,f) + \mathrm{Ric}(r,s,\Omega,f) - \mathrm{Ric}(r,s,\xi_a,f)$$

$$\leqq q\, \mathfrak{S} \log(T(r,s,) + c_2) + c_3$$

for all $0 < s < r$. (Griffiths-King [33] Lemma 7.30 and Carlson-Griffiths [8] Lemma 5.15.)

<u>Proof.</u> For $0 < s < r$ define

$$H(r,s) = \lambda\, T(r,s,L,\kappa,f) + \mathrm{Ric}(r,s,\Omega,f) - \mathrm{Ric}(r,s,\xi_a,f).$$

Then H is continuous. Observe $T(r,s,L,\kappa,f) = \sum_{j=1}^{q} T(r,s,L_j,\kappa_j,f)$. Now (17.7) states

$$H(r,s) = \sum_{j=1}^{q} \int_s^r \int_{M[t]} dd^c \mathfrak{L}(\delta u_j \circ f) \wedge \upsilon^{m-1} t^{1-2m}\, dt.$$

For $r \in \mathfrak{C}_\tau^0$ define

$$c(r) = \sum_{j=1}^{q} \tfrac{1}{2} \int_{M\langle r\rangle} \mathfrak{L}(\delta u_j \circ f)\, \sigma$$

If $0 < s < r \in \mathfrak{C}_\tau^0$ and $s \in \mathfrak{C}_\tau^0$, then (17.4) implies $H(r,s) = c(r) - c(s)$. By continuity, c extends to a continuous function on $\mathbb{R}^+$. Because $0 \leqq \delta u_j e < 1$, the function c is non-negative. Define $c_1 = -c(s) \leqq 0$. Then $c_1 \leqq H(r,s) \leqq c(r)$ for all $r > s$. Take $r \in \mathfrak{C}_\tau^0$. Define

$$c_2 = \sum_{j=1}^{q} m_{a_j}(s,L_j,\kappa_j,f) + \frac{q\mathfrak{S}}{2} |\log \delta| > 0.$$

Then

$$c(r) = \varsigma \sum_{j=1}^{q} \frac{1}{\varsigma} \int_{M\langle r\rangle} \log | \log (\delta u_j \circ f) | \, \sigma$$

$$\leqq \varsigma q \sum_{j=1}^{q} \frac{1}{q} \log \left(\frac{1}{\varsigma} \int_{M\langle r\rangle} |\log (\delta u_j \circ f)| \, \sigma \right)$$

$$\leqq \varsigma q \log \left(\sum_{j=1}^{q} \frac{1}{q\varsigma} \int_{M\langle r\rangle} \left(\log \frac{1}{\delta u_j \circ f}\right) \sigma \right)$$

$$= \varsigma q \log \left(\frac{2}{q\varsigma} \sum_{j=1}^{q} \left(m_{a_j}(r, L_j, \kappa_j, f) - \frac{\varsigma}{2} \log \delta\right)\right)$$

$$\leqq \varsigma q \log \left[\sum_{j=1}^{q} T(r, s, L_j, \kappa_j, f) + c_2 \right] + c_3 .$$

Hence

$$c_1 \leqq H(r,s) \leqq c(r) \leqq \varsigma q \log [T(r,s,L,\kappa,f) + c_2] + c_3$$

q.e.d.

c) The Third Main Theorem for several bundles

Let L_1 and L_2 be holomorphic line bundles on the compact, complex space X. Assume L_1 is positive. Then an integer $m > 0$ exists such that $L_1^m \otimes L_2^*$ is non-negative. Let $[L_2/L_1]$ the minimum of all these integers.

Theorem 18.13. (Third Main Theorem) Assume (C1) - (C5), (C8), (C9), (C11) and (C12). Assume $c(L,\kappa) > 0$. Take $w \in \mathbb{N}$. Assume $p = [(K_N^*)^w/L] < w$. Take $\varepsilon > 0$. Define

(18.17) $\qquad c_2 = (1+\frac{\varepsilon}{2})\, \varsigma\, n \qquad\qquad c_3 = 2\varepsilon\, \varsigma\, n$

Take $s > 0$. Then $T(r,s,L,\kappa,f) \longrightarrow \infty$ for $r \longrightarrow \infty$ and a constant $c_1 > 0$ exists such that

$$N_{\theta_F}(r,s) + (1 - \frac{p}{w})\, T(r,s,L,\kappa,f)$$

(18.18) $\|_\varepsilon$
$$\leqq \sum_{j=1}^{q} N_{a_j}(r,s,L_j,f) + \mathrm{Ric}_\tau(r,s)$$

$$+ c_1 \log T(r,s,L,\kappa,f) + c_2 \log Y(r) + c_3 \log r .$$

<u>Proof.</u> A hermitian metric $\hat{\kappa}$ along the fibers of $L^p \otimes K_N^w$ exists such that $c(L^p \otimes K_N^w, \hat{\kappa}) \geqq 0$. A hermitian metric $\tilde{\kappa}$ along the fibers of K_N exists such that $\hat{\kappa} = \kappa^p \otimes \tilde{\kappa}^w$. A form $\Omega > 0$ of class C^∞ and degree 2n exists such that $c(K_N, \tilde{\kappa}) = \mathrm{Ric}\,\Omega$.
Now $p < w$ implies

$$c(L,\kappa) + \mathrm{Ric}\,\Omega > \frac{1}{w}\, c(L^p \otimes K_N^w, \hat{\kappa}) \geqq 0.$$

Hence (C6) holds with $\lambda = 1$. By Theorem 16.3 a Carlson-Griffiths form exists satisfying (C7) for $\lambda = 1$. Since $a_1,\ldots,a_q$ are in general position, $E_{L_j}[a_j]$ is thin for $j = 1,\ldots,q$. By Lemma 18.1 and (C12) the map f has rank n. Since M is connected, $f^{-1}(E_{L_j}[a_j])$ is thin. Hence f is adapted by a_j for L_j for $j = 1,\ldots q$. Therefore (C1) - (C12) are satisfied with $\lambda = 1$ and with $\delta > 0$ and $\gamma > 0$ as determined by (16.5)-(16.9).

Abbreviate $T(r,s) = T(r,s,L,\kappa,f)$. A constant $\gamma_1 > 0$ exists such that $-\gamma_1 c(L,\kappa) \leqq \mathrm{Ric}\,\Omega \leqq \gamma_1 c(L,\kappa)$ holds on N, which implies

(18.19)
$$-\gamma_1 T(r,s) \leqq \mathrm{Ric}(r,s,\Omega,f) \leqq \gamma_1\, T(r,s)$$

for all $r > s > 0$. Also $p\, c(L,\kappa) + w\, \mathrm{Ric}\,\Omega \geqq 0$ implies

(18.20)
$$p\, T(r,s) + w\, \mathrm{Ric}(r,s,\Omega,f) \geqq 0.$$

By Proposition 18.12 constants $\gamma_2 > 0, \gamma_3 > 0$ and γ_4 exist such that

$$-\gamma_2 \leqq T(r,s) + \mathrm{Ric}(r,s,\Omega,f) - \mathrm{Ric}(r,s,\xi_a,f)$$

(18.21)

$$\leqq q\, \varsigma \log (T(r,s) + \gamma_3) + \gamma_4$$

Therefore

$$(18.22)\qquad 0 \leqq \mathrm{Ric}(r,s,\xi_a,f) \leqq (1+\gamma_1)\, T(r,s) + \gamma_2$$

Now, (18.9) implies $T(r,s) \longrightarrow \infty$ for $r \longrightarrow \infty$. Also

$$(18.23)\qquad \begin{aligned} (1-\tfrac{p}{w})\, T(r,s) &\leqq T(r,s) + \mathrm{Ric}(r,s,\Omega,f) \\ &\leqq \mathrm{Ric}(r,s,\xi_a,f) + q\, \mathfrak{S} \log\, (T(r,s)+\gamma_3) + \gamma_4. \end{aligned}$$

With (17.2) this leads to

$$(18.24)\qquad \begin{aligned} & N_{\theta_F}(r,s) + (1-\tfrac{p}{w})\, T(r,s) \\ & \leqq \sum_{j=1}^{q} N_{a_j}(r,s,L_j,f) + \mathrm{Ric}_\tau(r,s) + \mu_a(r) - \mu_a(s) \\ & \quad + q\, \mathfrak{S} \log\, (T(r,s)+\gamma_3) + \gamma_4. \end{aligned}$$

A constant $\gamma_5 > 1$ exists such that $(1+\gamma_1)\, T(r,s) + \gamma_2 \leqq \gamma_5\, T(r,s)$ for $r > s + 1$. Hence (18.22) and (18.15) imply

$$(18.25)\qquad \|_\varepsilon \quad \mu_a(r) \leqq c_2 \log T(r,s) + c_2 \log Y(r) + c_3 \log r + \gamma_6$$

where $\gamma_6 = c_2 \log \gamma_5$ is a constant. A constant $c_1 > 0$ exists such that

$$(18.26)\qquad \begin{aligned} & \gamma_6 + c_2 \log T(r,s) + q\, \mathfrak{S} \log\, (T(r,s)+\gamma_3) + \gamma_4 - \mu_a(s) \\ & \leqq c_1 \log T(r,s). \end{aligned}$$

Now, (18.24), (18.25) and (18.26) imply (18.18); q.e.d.

If τ is obtained by a covering map, certain simplifications take place and it is worthwhile to state them here.

Theorem 18.14. Assume (C1) - (C6). Let M be a connected complex manifold of dimension m. Let $f: M \longrightarrow N$ be a holomorphic map of rank n. Let $\beta: M \longrightarrow \mathbb{C}^m$ be a proper, surjective holomorphic map and let $\nu = \theta_{D\beta}$ be the branching divisor of β. Then $\tau = |\beta|^2$ is a parabolic exhaustion of M. Take $w \in \mathbb{N}$. Assume $p = [(K_N^*)^w/L] < w$. Take $\varepsilon > 0$ and $s > 0$. Then $T(r,s,L,\kappa,f) \longrightarrow \infty$ for $r \longrightarrow \infty$ and a constant $c > 0$ exists such that

$$(1-\tfrac{p}{w})\, T(r,s,L,\kappa,f) \leqq \sum_{j=1}^{q} N_{a_j}(r,s,L_j,f) + N_\nu(r,s)$$

(18.27) $\quad \|_\varepsilon$

$$+ c \log T(r,s,L,\kappa,f) + 2\,\varepsilon \varsigma\, n \log r .$$

Proof. Since β is surjective and M is connected, β has pure rank m. Hence τ is a parabolic exhaustion of M and (C8) holds. Since $a_1,\ldots a_q$ are in general position, each $E_{L_j}[a_j]$ is thin. Because f has rank n and because M is connected, $f^{-1}(E_{L_j}[a_j])$ is thin. Hence (C10) is satisfied. By Proposition 18.6 a holomorphic form φ of degree $m - n \geqq 0$ exists such that φ is effective for f and dominated by τ with a constant dominator $Y \equiv m$. Then φ induces an effective Jacobian sectic F_φ of f dominated by τ with the dominator $Y \equiv m$. Hence (C11) and (C12) are satisfied. Theorem 18.13 applies with $\mathrm{Ric}_\tau(r,s) = N_\nu(r,s)$ by Theorem 15.4. A constant $c > 0$ exists such that $c_1 \log T(r,s) + c_2 \log m \leqq c \log T(r,s)$ for $r > s + 1$. Hence (18.18) implies (18.28); q.e.d. (Observe that F_φ depends on choices, hence the valence function of the divisor of F_φ was dropped).

Also the Third Main Theorem is implicitly contained in the Carlson-Griffiths-King theory. An explicit formulation cannot be found in [33] or [8]. The Third Main Theorem corresponds to Nevanlinna's Second Main Theorem [49].

d) The Third Main Theorem for a single bundle

Although not strictly necessary but useful, $L_1 = \ldots = L_q$ is assumed. The following General Assumptions are made:

(D1) Let N be a compact, connected, complex manifold of dimension $n > 0$

(D2) Let L be a positive, holomorphic line bundle over N.

(D3) Let V be a linear subspace of $\Gamma(N,L)$ with $0 < \dim V = k + 1$. Let ℓ be a Hermitian metric on V. Let $\eta: N \times V \longrightarrow L$ be the evaluation map.

(D4) Let κ be a hermitian metric along the fibers of L such that $c(L,\kappa) > 0$. Assume κ is distinguished for η and ℓ.

(D5) Given are $a_1,\ldots,a_q$ in $\mathbb{P}(V)$ such that $a_1,\ldots,a_q$ are in general position for L.

(D6) Let M be a connected, complex manifold of dimension $m > 0$.

(D7) Let $f\colon M \longrightarrow N$ be a holomorphic map.

(D8) Let τ be a parabolic exhaustion of M.

(D9) Let F be an effective Jacobian section of f dominated by τ. Let Y be the dominator.

Remarks. By Kodaira and (D4), N is projective algebraic. (D5) implies that η is a semi-amplification. If $q > n$, then η is an amplification. (D9) implies $m \geqq n = \operatorname{rank} f$. Hence f is adapted to every $b \in \mathbb{P}(V)$ for L. Since $f^{-1}(E_L[\infty]) \subseteq f^{-1}(E_L[b])$ for all $b \in \mathbb{P}(V)$, the map f is safe for L. In fact f is almost adapted of order 0 for L. Define $\eta^q = \eta \otimes\ldots\otimes \eta$. Then η^q is a semi-amplification of L^q and L^q is adapted to $a = a_1\otimes\ldots\otimes a_q \in \mathbb{P}(\otimes_q V)$. Assume (D1) - (D9). Then (C1) - (C5), (C9) - (C12) are satisfied using the translation table.

C	N	n	L_j	κ_j	q	V_j	η_j	a_j	L	κ	V	a	η	M	m	C	τ	f	F	Y
D	N	n	L	κ	q	V	η	a_j	L^q	κ^q	$\otimes_q V$	a	η^q	M	m	D	τ	f	F	Y

Also $c(L^q,\kappa^q) = qc(L,\kappa)$ and $T(r,s,L^q,f) = q\,T(r,s,L,\kappa,f)$. Abbreviate $T(r,s) = T(r,s,L,\kappa,f)$. Then Theorems 18.13 and 18.14 translate to:

Theorem 18.13E. (Third Main Theorem). Assume (D1) - (D9). Take $w \in \mathbb{N}$. Define $p = [(K_N^*)^w/L]/w$. Assume $q > p$. Take $\varepsilon > 0$. Define c_2 and c_3 by (18.17). Take $s > 0$. Then $T(r,s) \longrightarrow \infty$ for $r \longrightarrow \infty$ and a constant $c_1 > 0$ exists such that

$$(18.28)\|_\varepsilon \qquad \begin{aligned} &N_{\theta_F}(r,s) + (q-p)\,T(r,s) \\ &\leqq \sum_{j=1}^{q} N_{a_j}(r,s) + \operatorname{Ric}_\tau(r,s) + c_1 \log T(r,s) \\ &\quad + c_2 \log Y(r) + c_3 \log r. \end{aligned}$$

(Observe, here c_1 has to be enlarged, to switch from log (qT) to log T. Also

$$[(K_N^*)^{wq}/L^q] = [(K_N^*)^w/L] = pw < qw.$$

Hence the p in Theorem 18.13 is the pw in Theorem 18.13E and the w in Theorem 18.13 is the qw in Theorem 18.13E).

Theorem 18.14E. Assume (D1) - (D7). Assume that f has rank n. Let $\beta: \longrightarrow \mathbb{C}^m$ be a proper, surjective, holomorphic map. Let $\nu = \theta_{D\beta}$ be the branching divisor of β. Then $\tau = |\beta|^2$ is a parabolic exhaustion of M. Take $\varepsilon > 0$ and $s > 0$. Then $T(r,s) \longrightarrow \infty$ for $r \longrightarrow \infty$. A constant $c > 0$ exists such that

$$(q-p)\ T(r,s) \leqq \sum_{j=1}^{q} N_{a_j}(r,s) + N_\nu(r,s)$$

(18.29) $\|_\varepsilon$

$$+ c \log T(r,s) + 2\ \varepsilon\, c_n \log r.$$

Let L_1 and L_2 be holomorphic line bundles on N. Assume L_1 is positive. Then the infimum $[L_2:L_1]$ of all quotients v/w exists where $v \geqq 0$ and $w > 0$ are integers such that $L_1^v \otimes (L_2^*)^w$ is non-negative. Obviously

(18.30) $$[L_2:L_1] = \inf \{[L_2^w/L_1]/w \mid w \in \mathbb{N}\}.$$

Because the constants c_1 in (18.28) and c in (18.29) depend on w, the number p cannot be replaced by $[K_N^*:L]$.

Compare the results of this section with Griffiths-King [33] 7, Carlson-Griffiths [8] 5, and Drouilhet [20] Proposition 3.23. See also Weyl [89], Stoll [67], Murray [48] and Wong [90].

19. The Defect Relation

Assume (D1) - (D9). Take $s > 0$. Then $T(r,s) \longrightarrow \infty$ for $r \longrightarrow \infty$. The Ricci defect of f is defined by

(19.1)
$$R_f = \limsup_{r\to\infty} \frac{\mathrm{Ric}_\tau(r,s)}{T(r,s)}.$$

The dominator defect of f is defined by

(19.2)
$$Y_F = \limsup_{r\to\infty} \frac{\log Y(r)}{T(r,s)} \geqq 0.$$

The ramification defect of f is defined by

(19.3)
$$\Theta_F = \liminf_{r\to\infty} \frac{N_{\theta_F}(r,s)}{T(r,s)}.$$

The (Nevanlinna) defect of f for $a \in \mathbb{P}(V)$ and L is defined by

(19.4)
$$\delta_f(a,L) = \liminf_{r\to\infty} \frac{m_a(r)}{T(r,s)}$$

Since $T(r,s) - T(r,s')$ is constant, and since (12.39) holds, the defects R_f, Y_F, Θ_F and $\delta_f(a,L)$ do not depend on s and the hermitian metric κ. The First Main Theorem shows

(19.5)
$$1 - \delta_f(a,L) = \limsup_{r\to\infty} \frac{N_a(r,s)}{T(r,s)}$$

(19.6)
$$0 \leqq \delta_f(a,L) \leqq 1$$

Especially $\delta_f(a,L) = 1$ if $f(M) \cap E_L[a] = \emptyset$.

Theorem 19.1 (Defect Relation). Assume (D1) - (D9). Assume $q > [K_N^*:L]$. Then

$$\Theta_F + \sum_{j=1}^{q} \delta_f(a_j,f) \leqq [K_N^*:L] + R_f + c_n Y_F.$$

Proof. Observe $T(r,s)/\log r \longrightarrow A(\infty) > 0$ for $r \longrightarrow \infty$. Hence Theorem 18.13E and 18.17 with $\varepsilon \longrightarrow 0$ imply Theorem 19.1; q.e.d.

The holomorphic map f is said to have transcendental growth (respectively rational growth), if $A(\infty) = \infty$ (respectively if $A(\infty) < \infty$). Observe

$$(19.7) \qquad \frac{T(r,s)}{\log r} \longrightarrow A(\infty) \qquad \text{for } r \longrightarrow \infty.$$

The concept of rational and transcendental growth of a divisor was introduced in (11.15). The following Lemma will be needed.

Lemma 19.2. Assume (D6). Let $\beta : M \longrightarrow \mathbb{C}^m$ be a proper, surjective holomorphic map. Then $\tau = |\beta|^2$ is a parabolic exhaustion of M. Let $\nu \geqq 0$ be a holomorphic divisor on M. Define $S = \operatorname{supp} \nu$. Then $\beta(S)$ is analytic and thin in M. Define

$$(19.8) \qquad B = \{\mathfrak{z} \in \beta(S) \mid \dim_{\mathfrak{z}} \beta(S) = m - 1\}.$$

Then ν has rational growth, if and only if B is affine algebraic.

Proof. Since B is proper, $C = \beta(S)$ is analytic in $\mathbb{C}^m$ with $\dim C \leqq m - 1$. Because β is surjective, $E = \{x \in M \mid \operatorname{rank}_x \beta < m\}$ is thin analytic in M. Also $D = \beta(E)$ is thin analytic in $\mathbb{C}^m$ with $\dim D \leqq m - 2$ ([2], Proposition 1.24). Hence $\tilde{D} = \beta^{-1}(D)$ is thin analytic in M. Define $M_0 = M - \tilde{D}$ and $G = \mathbb{C}^m - D$. Then $\beta_0 = \beta : M_0 \to G$ is a proper, surjective light holomorphic map of finite sheet number s. Define $S_0 = S \cap M_0$ and $C_0 = C \cap G$. Then $\gamma_0 = \beta_0 : S_0 \longrightarrow C_0$ is a proper, surjective, light holomorphic map with $\# \gamma_0^{-1}(\mathfrak{z}) \leqq s$ for all $\mathfrak{z} \in C_0$. Then $S_1 = \overline{S}_0$ and $C_1 = \overline{C}_0$ are analytic subsets of M respectively $\mathbb{C}^m$ with $S_1 \subseteq S$ and $C_1 \subseteq C \subseteq C_1 \cup D$. Since γ_0 is light, C_1 has pure dimension m - 1. Hence $C_1 = B$.

Let H be a branch of B with $H \subseteq \tilde{D}$. Let $j: H \longrightarrow M$ and $\iota : D \longrightarrow \mathbb{C}^m$ be the inclusions. Define $\gamma_1 = \beta : H \longrightarrow D$. Then $\iota \circ \gamma_1 = \beta \circ j$. Define $\hat{\tau}(\mathfrak{z}) = |\mathfrak{z}|^2$ and $\hat{\upsilon} = dd^c \hat{\tau}$. Then $\upsilon = \beta^*(\hat{\upsilon})$ and $\iota^*(\hat{\upsilon}^{m-1}) = 0$ since $\dim D \leqq m - 2$. Hence

$$j^*(\upsilon^{m-1}) = j^*(\beta^*(\upsilon^{m-1})) = \gamma_1^*\iota^*(\hat{\upsilon}^{m-1})) = 0.$$

Therefore

$$r^{2m-2}n_\nu(r) = \int_{S[r]} \nu\upsilon^{m-1} = \int_{S_1[r]} \nu\upsilon^{m-1} = \int_{S_0[r]} \nu\upsilon^{m-1}.$$

Assume ν has rational growth. Then $n_\nu(r) \leqq n_\nu(\infty) = K < \infty$. Hence

$$r^{2m-2}K \geqq \int_{S_0[r]} \upsilon^{m-1} \geqq \int_{C_0[r]} \hat{\upsilon}^{m-1} = r^{2m-2}n_B(r).$$

Hence $K \geqq n_B(\infty)$. According to [68] or to Rutishauser [59]§13, B is affine algebraic.

Assume B is affine algebraic. Then $B = C_1 = \overline{C}_0$ has only finitely many branches. Therefore C_0 has only finitely many branches. Because $\gamma_0 \colon S_0 \longrightarrow C_0$ is proper, light, surjective and holomorphic, S_0 has only finitely many branches. Hence ν is bounded on $\mathfrak{R}(S_0)$. Let K_0 be a bound. Then

$$n_\nu(r) \leqq \frac{K_0}{r^{2m-2}} \int_{S_0[r]} \nu^{m-1} \leqq \frac{sK_0}{r^{2m-2}} \int_{C_0[r]} \hat{\upsilon}^{m-1} = sK_0 n_B(r).$$

Therefore $n_\nu(r) \leqq sK_0 n_B(\infty) < \infty$; q.e.d.

If β is not light, ν may have rational growth (even $n_\nu(r) \equiv 0$ may occur) but $\beta(\operatorname{supp} \nu)$ may be transcendental. For instance, let ν be the branching divisor of a modification blowing up $\mathbb{Z}^m$ in $\mathbb{C}^m$.

Theorem 19.3. Assume (D1) - (D7). Assume f has rank n. Let $\beta \colon M \longrightarrow \mathbb{C}^m$ be a proper, holomorphic map, and let $\nu = \theta_{D\beta}$ be the branching divisor of β. Denote $S = \operatorname{supp} \nu$. Define B by (19.8). Then $\tau = |\beta|^2$ is a parabolic exhaustion of M. Then $T(r,s) \longrightarrow \infty$ for $r \longrightarrow \infty$ and

(19.9) $$R_f = \limsup_{r\to\infty} \frac{N_\nu(r,s)}{T(r,s)} \geqq 0$$

$$(19.10)\qquad \sum_{j=1}^{q} \delta_f(a_j,L) \leqq [K_N^*:L] + R_f$$

If B is affine algebraic and if f has transcendental growth, then

$$(19.11)\qquad \sum_{j=1}^{q} \delta_f(a_j,L) \leqq [K_N^*:L].$$

Proof. Since $\mathrm{Ric}_\tau(r,s) = N_\nu(r,s)$, the identity (19.9) follows from (19.1). Especially $R_f \geqq 0$. Hence (19.10) and (19.11) are trivial if $q \leqq [K_N^*:L]$. Assume $q > [K_N^*:L]$. Since $T(r,s)/\log r \longrightarrow A(\infty) > 0$ the estimates (19.10) follows from Theorem 18.14.E. If B is affine algebraic then $N_\nu(r,s)/\log r \longrightarrow n_\nu(\infty)$ for $r \longrightarrow \infty$ with $n_\nu(\infty) < \infty$. If f has transcendental growth then $T(r,s)/\log r \longrightarrow \infty$ for $r \longrightarrow \infty$. Hence

$$R_f = \limsup_{r\to\infty} \frac{N_\nu(r,s)}{\log r}\,\frac{\log r}{T(r,s)} = 0$$

q.e.d.

If β is biholomorphic, then B is empty and (19.11) holds. Hence (19.11) is valid if $M = \mathbb{C}^m$ and $\tau(\mathfrak{z}) = |\mathfrak{z}|^2$ for all $\mathfrak{z} \in \mathbb{C}^m$. If in the case of Theorem 19.1 or 19.3 either $R_f = +\infty$ or $Y_F = +\infty$, the statement is useless. If however $R_f = Y_F = 0$, then $f(M) \cap E_L[a_j] \neq \emptyset$ for at least one $j \in \mathbb{N}[1,q]$ if $[K_N^*:L] < q$ is assumed. If $N = \mathbb{P}(W)$ with $W = V^*$ and if $L = S_0(W)^*$ is the hyperplane section bundle, then $E_L[a_j] = \ddot{E}[a_j]$ and $[K_N^*:L] = n + 1$. The Third Main Theorems and the Defect Relations for maps into projective space are obtained. For more details see [82] §21. Compare these results with Griffiths-King [33] 7 and Carlson-Griffiths [8] 5. See also Weyl [89] and Stoll [67].

The Casorati-Weierstrass Theorems involves the defect $\delta_f^0(a,L)$ but the Defect Relation involves the defect $\delta_f(a,L)$. If η is ample, then $\delta_f^0(a,L) = \delta_f(a,L)$, but if η is only semi-ample the defects may be different. This shall be investigated now.

Assume (D1) - (D4), (D6) - (D9). Assume also that f is safe of order 1. i.e. $\dim f^{-1}(E_L[\infty]) \leqq m-1$. Because N is connected and $V \subseteq \Gamma(N,L)$, the set $E_L[a]$ is thin for each $a \in \mathbb{P}(V)$. Since f has rank n by (D9), also $f^{-1}(E_L[a])$ is thin. Hence f is adapted to all $a \in \mathbb{P}(V)$ for L. Also (B1) - (B12) holds with $p = 1$. By Theorem 13.1, $A[r] > 0$

or all $r > 0$ and $T[r,s] \longrightarrow \infty$ for $r \longrightarrow \infty$; also $T[r,s]/\log r \to A[\infty] > 0$ for $r \to \infty$. Hence the following defects do not depend on s:

$$R_f^0 = \limsup_{r\to\infty} \frac{Ric_\tau(r,s)}{T[r,s]} \tag{19.12}$$

$$Y_F^0 = \limsup_{r\to\infty} \frac{\log Y(r)}{T[r,s]} \geqq 0 \tag{19.13}$$

$$\Theta_F^0 = \limsup_{r\to\infty} \frac{N_{\theta_F}(r,s)}{T[r,s]} \geqq 0 \tag{19.14}$$

$$\delta_f^0(a,L) = \liminf_{r\to\infty} \frac{m_a[r]}{T[r,s]} = 1 - \limsup_{t\to\infty} \frac{N_a(r,s)}{T[r,s]} \tag{19.15}$$

$$0 \leqq \delta_f^0(a,L) \leqq 1. \tag{19.16}$$

ake $r \in \mathfrak{C}_\tau$. Then (4.3), (12.26) and (12.27) show that

$$\Xi(r) = \int_{M\langle r\rangle} \log \frac{1}{\|\infty, f\|_\kappa} \sigma = m_a(r) - m_a[r]. \tag{19.17}$$

xists and is independent of $a \in \mathbb{P}(V)$. Also (19.17) shows that Ξ extends o a continuous function on $\mathbb{R}^+$. Since κ is distinguished, $\Xi \geqq 0$. The irst Main Theorem implies

$$T(r,s) - T[r,s] = \Xi(r) - \Xi(s) \tag{19.18}$$

efine the lower and upper <u>deviation defects</u> by

$$\Xi_\kappa = \liminf_{r\to\infty} \frac{\Xi(r)}{T(r,s)} \geqq 0 \tag{19.19}$$

$$\Xi^\kappa = \limsup_{r\to\infty} \frac{\Xi(r)}{T(r,s)} \geqq 0. \tag{19.20}$$

ecause $T(r,s) \geqq \Xi(r) - \Xi(s)$, the following estimate holds

$$0 \leqq \Xi_\kappa \leqq \Xi^\kappa \leqq 1 \tag{19.21}$$

lso (19.18) implies

$$(19.22)\qquad 0 \leqq 1 - \Xi_{\kappa} = \limsup_{r\to\infty} \frac{T[r,s]}{T(r,s)} \leqq 1$$

$$(19.23)\qquad 0 \leqq 1 - \Xi^{\kappa} = \liminf_{r\to\infty} \frac{T[r,s]}{T(r,s)} \leqq 1.$$

Therefore, if $\mathrm{Ric}_{\tau}(r,s) \geqq 0$ for all $r \geqq r_0$ for some $r_0 > 0$, then

$$(19.24)\qquad (1-\Xi^{\kappa})\, R_f^0 \leqq R_f \leqq (1-\Xi_{\kappa})\, R_f^0$$

$$(19.25)\qquad (1-\Xi^{\kappa})\, Y_F^0 \leqq Y_F \leqq (1-\Xi_{\kappa})\, Y_F^0$$

$$(19.26)\qquad (1-\Xi^{\kappa})\, \Theta_F^0 \leqq \Theta_F \leqq (1-\Xi_{\kappa})\, \Theta_F^0$$

$$(19.27)\qquad (1-\Xi^{\kappa})(1-\delta_f^0) \leqq (1-\delta_f) \leqq (1-\Xi_{\kappa})(1-\delta_f^0)$$

$$(19.28)\qquad \Xi_{\kappa} + (1-\Xi_{\kappa})\, \delta_f^0 \leqq \delta_f \leqq \Xi^{\kappa} + (1-\Xi^{\kappa})\delta_f^0$$

Observe that δ_f^0 and δ_f depend on a but Ξ_{κ} and Ξ^{κ} do not. The map f does <u>not deviate</u> if $\Xi^{\kappa} = 0$. Then $\Xi_{\kappa} = 0 = \Xi^{\kappa}$ and $R_f^0 = R_f$, $Y_F^0 = Y_F$, $\Theta_F^0 = \Theta_F$ and $\delta_f^0(a,L) = \delta_f(a,L)$.

Actually these estimates are quite surprising. For instance if $\Xi_{\kappa} > 0$, then $\delta_f(a,L) > 0$ for all $a \in \mathbb{P}(V)$.

20. The algebraic case

a) Meromorphic maps

<u>Proposition 20.1.</u> Let S be a thin analytic subset of the complex space M. Let V be a complex vector space of dimension k + 1. Let $f: M - S \longrightarrow \mathbb{P}(V)$ be a meromorphic map. Then f extends to a meromorphic map $\hat{f}: M \longrightarrow N$ if and only if for every $a \in \mathbb{P}(V^*)$ the closure of $f^{-1}(\ddot{E}[a])$ is analytic in M.

<u>Proof.</u> Clearly the condition is necessary. Hence assume that $E_a = f^{-1}(\ddot{E}[a])$ is analytic for each $a \in \mathbb{P}(V^*)$. W.ℓ.o.g. assume that f is holomorphic (by enlarging M−S) and that M is irreducible. Take $a \in \mathbb{P}(V^*)$ such that $E_a \neq M$. Hence E_a is thin. Let $\alpha_0, \ldots, \alpha_k$ be a base

of V^* with $\mathbb{P}(\alpha_0) = a$. Let $\mathfrak{a}_0,\ldots,\mathfrak{a}_k$ be the dual base. Then $V_a = \mathfrak{a}_0 + \mathbb{C}\,\mathfrak{a}_1 + \ldots + \mathbb{C}\,\mathfrak{a}_k$ is a hyperplane in V. Let $j\colon V_a \longrightarrow V$ be the inclusion. The map $\mathbb{P}_a = \mathbb{P}\colon V_a \longrightarrow \mathbb{P}(V) - \ddot{E}[a]$ is biholomorphic. Define $M_a = M - (S \cup E_a)$. Then $\mathfrak{w} = j \circ \mathbb{P}_a^{-1} \circ f\colon M_a \longrightarrow V$ is holomorphic with $\mathbb{P} \circ \mathfrak{w} = f$ on M_a. Define $w_\mu = \alpha_\mu \circ \mathfrak{w}$. Then $\mathfrak{w} = \mathfrak{a}_0 + w_1\alpha_1 + \ldots + w_k\alpha_k$.

For each $b \in \mathbb{C}$, define $H_\lambda(b) = w_\lambda^{-1}(b)$ for $\lambda = 1,\ldots,k$. Take $\lambda \in \mathbb{N}[1,k]$ and $b \in \mathbb{C}$. Define $\gamma = \alpha_\lambda - b\alpha_0 \in V^*$. Then $\gamma \neq 0$ and $c = \mathbb{P}(\gamma) \in \mathbb{P}(V^*)$ with

$$E_c \cap M_a = f^{-1}(\ddot{E}[c]) - E_a = \mathfrak{w}^{-1}(E[c])$$

$$= (\gamma \circ \mathfrak{w})^{-1}(0) = w_\lambda^{-1}(b) = H_\lambda(b).$$

Because $S \cup E_a$ is thin, $\overline{H_\lambda(b)}$ is analytic in M for every $b \in \mathbb{C}$. By Thullen [70], w_λ extends to a meromorphic function on M. Take $x_0 \in M$. A holomorphic function $h\colon U \longrightarrow \mathbb{C}$ on an open neighborhood U of x_0 exists such that $h^{-1}(0)$ is thin and such that $\mathfrak{v} = h\,\mathfrak{w}$ is a holomorphic vector function on U. Then $\mathbb{P} \circ \omega = \mathbb{P} \circ \mathfrak{w} = f$. Hence $\mathfrak{v}$ is a representation of f on U. Therefore f extends to a meromorphic map $\hat{f}\colon M \longrightarrow \mathbb{P}(V)$; q.e.d.

Lemma 20.2. Let S be a thin analytic subset of the complex space M. Let N and P compact complex spaces. Let $\varphi\colon N \longrightarrow P$ be an injective holomorphic map. Let $f\colon M - S \longrightarrow N$ be a meromorphic map. Then f is meromorphic on M if and only $\varphi \circ f$ is meromorphic on M.

The proof is left to the reader or see [82] Lemma 22.2.

b) Projective closure of a vector space

Let W be a complex vector space of dimension $k > 0$. Then $\overline{\overline{W}} = \mathbb{P}(W \oplus \mathbb{C})$ is called the projective closure of W. If $a \in \mathbb{C}$, define $\tilde{j}_a\colon W \longrightarrow W \oplus \mathbb{C}$ by $\tilde{j}_a(\mathfrak{w}) = (\mathfrak{w}, a)$. Define $j = \mathbb{P} \circ \tilde{j}_1$. Then $j(W)$ is open and dense in $\overline{\overline{W}}$ and $\overline{\overline{W}} - j(W)$ is a hyperplane. The map $j\colon W \longrightarrow j(W)$ is biholomorphic. Identify W and $j(W)$ by j such that j becomes the inclusion map $W \subseteq \overline{\overline{W}}$. An injective holomorphic map $\iota\colon \mathbb{P}(W) \longrightarrow \overline{\overline{W}}$ is uniquely defined by $\iota \circ \mathbb{P} = \mathbb{P} \circ \tilde{j}_0$ on $W - \{0\}$. The map $\iota\colon \mathbb{P}(W) \longrightarrow \iota(\mathbb{P}(W))$ is biholomorphic. Identify $\mathbb{P}(W)$ and $\iota(\mathbb{P}(W))$ by ι such that ι becomes the inclusion map $\mathbb{P}(W) \subseteq \overline{\overline{W}}$. Then $\overline{\overline{W}} = W \cup \mathbb{P}(W)$ is a disjoint union. With these identification $\mathbb{P}(\mathfrak{y},0) = \mathbb{P}(\mathfrak{y})$ if $0 \neq \mathfrak{y} \in W$ and $\mathbb{P}(\mathfrak{y},z) = \mathfrak{y}/z$ if $\mathfrak{y} \in W$ and $0 \neq z \in \mathbb{C}$. Let $\pi: W \oplus \mathbb{C} \to W$ be the projection. Then $\ker \pi = \{0\} \oplus \mathbb{C}$. Hence one and only one holomorphic map $\overline{\overline{\mathbb{P}}}: \overline{\overline{W}} - \{0\} \longrightarrow \mathbb{P}(W)$ exists such that $\overline{\overline{\mathbb{P}}} \circ \mathbb{P} = \mathbb{P} \circ \pi$ on $(W-\{0\}) \oplus \mathbb{C}$ as shown in the diagram

$$\begin{array}{ccc} (W-\{0\}) \oplus \mathbb{C} & \xrightarrow{\quad\pi\quad} & W - \{0\} \\ \Big\downarrow \mathbb{P} & & \Big\downarrow \mathbb{P} \\ \overline{\overline{W}} - \{0\} & \xrightarrow{\quad\overline{\overline{\mathbb{P}}}\quad} & \mathbb{P}(W) \end{array}$$

Then $\overline{\overline{\mathbb{P}}}\,|\,(W-\{0\}) = \mathbb{P}$ and $\overline{\overline{\mathbb{P}}}\,|\,\mathbb{P}(W) = \text{Identity}$. If $W = \mathbb{C}^m$, then

$$\overline{\overline{\mathbb{C}}}^m = \mathbb{P}_m = \mathbb{P}(\mathbb{C}^{m-1}) = \mathbb{C}^m \cup \mathbb{P}_{m-1}.$$

Let Y be a linear subspace of W. Then $Y \oplus \mathbb{C}$ is a linear subspace of $W \oplus \mathbb{C}$ and $\overline{\overline{Y}} = \mathbb{P}(Y \oplus \mathbb{C})$ is a projective plane in $\overline{\overline{W}}$ with $Y = \overline{\overline{Y}} \cap W$ and $\mathbb{P}(Y) = \overline{\overline{Y}} \cap \mathbb{P}(W)$. Also $\overline{\overline{Y}}$ is the closure of Y in $\overline{\overline{W}}$. An <u>affine plane</u> A parallel to Y is any coset of W/Y. If $\mathfrak{a} \in A$, then $A = Y + \mathfrak{a}$. Let E be the linear hull in $W \oplus \mathbb{C}$ spanned by $A \oplus \{1\}$ and $Y \oplus \{0\}$. Then $\mathbb{P}(E)$ is a projective plane in $\overline{\overline{W}}$ with $A = \mathbb{P}(E) \cap W$ and $\mathbb{P}(Y) = \mathbb{P}(E) \cap \mathbb{P}(W)$ and $\overline{A} = \mathbb{P}(E) = A \cup \mathbb{P}(Y)$, where $\overline{A}$ is the closure of A in $\overline{\overline{W}}$. If $E \neq W \oplus \{0\}$ is a linear subspace of $W \oplus \mathbb{C}$, then $E \cap (W \oplus \{0\}) = Y \oplus \{0\}$ and $E \cap (W \oplus \{1\}) = A \oplus \{1\}$ where A is an affine plane parallel to Y.

Let Z be a complex vector space. Let $\alpha = W \longrightarrow Z$ be linear. Define $X = \ker \alpha$. A holomorphic map $\alpha^0: \mathbb{P}(W) - \mathbb{P}(X) \longrightarrow \mathbb{P}(Z)$ exists uniquely such that $\mathbb{P} \circ \alpha = \alpha^0 \circ \mathbb{P}$ on $W - X$. The linear map $\alpha \oplus \mathrm{Id}: W \oplus \mathbb{C} \longrightarrow Z \oplus \mathbb{C}$ has kernel $X \oplus \{0\}$. Hence

$$\tilde{\alpha} = (\alpha \oplus \mathrm{Id})^0: \overline{\overline{W}} - \mathbb{P}(X) \longrightarrow \overline{\overline{Z}} \tag{20.1}$$

is holomorphic with $\tilde{\alpha}|W = \alpha$ and $\tilde{\alpha}\,|\,(\mathbb{P}(W) - \mathbb{P}(X)) = \alpha^0$. The maps $\tilde{\alpha}$ and $\tilde{\alpha}^0$ are meromorphic on $\overline{\overline{W}}$ respectively $\mathbb{P}(W)$. If α is injective, then $\mathbb{P}(X) = \emptyset$ and $\tilde{\alpha}$ and α^0 are holomorphic on $\overline{\overline{W}}$ respectively $\mathbb{P}(W)$.

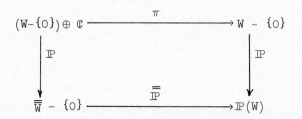

Let S be a projective plane of dimension s in $\mathbb{P}(W)$. Let Y be a linear subspace of dimension k - s - 1 in W such that $S \cap \overline{\overline{Y}} = \emptyset$. A holomorphic map

$$(20.2) \qquad \gamma = \gamma_Y^S : \overline{\overline{W}} - S \longrightarrow \overline{\overline{Y}}$$

called the <u>projection from center S onto $\overline{\overline{Y}}$</u> shall be defined:

One and only one linear subspace X of W exists with $\mathbb{P}(X) = S$. Then $W = X \oplus Y$. Let $\alpha: W \longrightarrow Y$ be the projection. Define $\gamma = \tilde{\alpha}$. Another description of γ can be given: Take $w \in \overline{\overline{W}} - S$, a projective plane Q(w) of dimension s + 1 exists such that $Q(w) \supseteq \{w\} \cup S$. Then $\{\gamma(w)\} = Q(w) \cap \overline{\overline{Y}}$. Also

$$(20.3) \qquad \gamma^{-1}(\gamma(w)) = Q(w) - S \approx \mathbb{C}^{s+1}.$$

Assume a hermitian metric is given on W. Define $\tau_W(\mathfrak{z}) = |\mathfrak{z}|^2$. Then $\upsilon_W = dd^c\tau_W$ and $\omega_W = dd^c \log \tau_W$ and the Fubini Study form $\ddot{\omega}_W$ on $\mathbb{P}(W)$ are defined with $\omega_W = \mathbb{P}^*(\ddot{\omega}_W)$. Since $\mathbb{P}: W - \{0\} \longrightarrow \mathbb{P}(W)$ extends holomorphically to $\overline{\mathbb{P}}: \overline{\overline{W}} - \{0\} \longrightarrow \mathbb{P}(W)$, also ω_W extends as a real analytic form to $\overline{\overline{W}} - \{0\}$ by $\omega_W = \overline{\overline{\mathbb{P}}}^*(\ddot{\omega}_W)$. Also $\mathbb{P} \circ \iota = \mathrm{Id}$ implies $\iota^*(\omega_W) = \ddot{\omega}_W$ on $\mathbb{P}(W)$.

An analytic subset N of $\mathbb{P}(W)$ is said to be a <u>projective algebraic variety embedded into $\mathbb{P}(W)$</u>. A complex space N is said to be a <u>projective algebraic variety</u> if there exists a biholomorphic map $\beta: N \longrightarrow N'$ onto a projective algebraic variety N' embedded into some projective space. An analytic subset M of W is said to be an <u>affine algebraic variety embedded into W</u> if the closure $\overline{M}$ of M in $\overline{\overline{W}}$ is analytic, which is the case if and only if M is the common zero set of polynomials on W (Chow [13]). If M is pure p-dimensional, then M is affine algebraic if and only if $n_M(\infty) < \infty$ ([71]). A complex space M is said to be an <u>affine algebraic variety</u> if M is biholomorphically equivalent to an embedded affine algebraic variety. In all this the word "variety" is replaced by "manifold" if the space is a manifold. An analytic precycle ν on W is said to be affine algebraic, if and only if $\mathrm{supp}_* \nu$ is affine algebraic in W.

Let M be an affine algebraic variety embedded into W. Let N be a projective algebraic space. Let $f: M \longrightarrow N$ be a meromorphic map. Then f is said to be rational, if f extends to a meromorphic map $\tilde{f}: \overline{M} \longrightarrow N$. If N is embedded, then $\tilde{f}$ is given by rational functions.

c) Affine exhaustion

The following General Assumptions shall be made.

(E1) Let W be a hermitian vector space of dimension $k > 0$. Define τ_W by $\tau_W(\mathfrak{z}) = |\mathfrak{z}|^2$.

(E2) Let M be a m-dimensional, irreducible, affine algebraic variety embedded into W with $0 < m < k$.

(E3) Let S be a projective plane of dimension $k - m - 1$ in $\mathbb{P}(W)$ such that $S \cap \overline{M} = \emptyset$ if $\overline{M}$ is the closure of M in $\overline{\overline{W}}$.

(E4) Let Y be a m-dimensional linear subspace of W such that $\overline{\overline{Y}} \cap S = \emptyset$. Let $\gamma_Y^S: (\overline{\overline{W}}-S) \longrightarrow \overline{\overline{Y}}$ be the projection from center S onto $\overline{\overline{Y}}$. Define $\tilde{\gamma} = \gamma_Y^S|\overline{M}$ and $\gamma = \tilde{\gamma}|M$.

(E5) Define $\tau_Y = \tau_W|Y$ and $\tau = \tau_Y \circ \gamma: M \longrightarrow \mathbb{R}_+$.

The maps $\tilde{\gamma}: \overline{M} \longrightarrow \overline{\overline{Y}}$ and $\gamma: M \longrightarrow Y$ are proper, surjective and holomorphic. If $y \in \overline{\overline{Y}}$, then $\tilde{\gamma}^{-1}(y)$ is a compact analytic subset of $(\gamma_Y^S)^{-1}(y) - S \approx \mathbb{C}^{k-m}$. Hence $\tilde{\gamma}$ and γ are light. Since τ_Y is the norm of a hermitian metric on Y, the function τ is a parabolic exhaustion of M. Also $\tilde{\omega} = \tilde{\gamma}^*\mathbb{P}^*(\ddot{\omega}_Y)$ is a real analytic non-negative form on $\overline{M} - M[0]$ with $\omega = \tilde{\omega}$ on $M - M[0]$.

Lemma 20.3. Assume (E1) - (E5). Let $\nu \geqq 0$ be an analytic precycle of dimension p in M. Then ν is affine algebraic in W if and only if $n_\nu(\infty) < \infty$.

Proof. Define $A = \operatorname{supp}_*\nu$. Assume ν is affine algebraic. Then A has only finitely many branches. Hence ν is bounded on $\mathfrak{R}(A)$. Let C be a bound. Then $n_\nu(\infty) \leqq C \int_{\overline{A}} \tilde{\omega}^p < \infty$. Now, assume $n_\nu(\infty) < \infty$. Because γ is light, $\gamma(A) = B$ is a pure p-dimensional analytic subset of Y with $n_\nu(\infty) \geqq n_B(\infty)$. Hence B is affine algebraic. Hence $\overline{B}$ is a pure p- dimensional analytic subset of $\overline{\overline{Y}}$. Then $D = \tilde{\gamma}^{1}(\overline{B})$ is a pure p-dimensional analytic subset of $\overline{M}$ with $D \supseteq \overline{A}$ where $D \cap (\overline{M}-M) = \tilde{\gamma}^{-1}(\overline{B}-B)$ has pure dimension $p - 1$. Hence $\overline{A}$ is analytic; q.e.d.

<u>Theorem 20.4.</u> Assume (E1) - (E5). Let M be smooth. Then the branching divisor $\nu = \theta_{D\gamma}$ of γ is affine algebraic and

$$(20.4) \qquad \mathrm{Ric}_\tau(r,s) = N_\nu(r,s) \qquad \lim_{r\to\infty} \frac{\mathrm{Ric}_\tau(r,s)}{\log r} = n_\nu(\infty) < \infty.$$

<u>Proof.</u> Let A be the largest open subset of $\overline{M}$ such that $\tilde{\gamma}|A$ is locally biholomorphic. Then $S = \overline{M} - A$ is analytic in $\overline{M}$ with $S \cap M = \mathrm{supp}\,\nu = \mathrm{supp}_*\nu$. Hence ν is affine algebraic and $n_\nu(\infty) < \infty$; q.e.d.

<u>Theorem 20.5.</u> Assume (E1) - (E5). Let L be a holomorphic line bundle on the complex space N. Let κ be a hermitian metric along the fibers of L with $c(L,\kappa) \geqq 0$. Let $f: M \longrightarrow N$ be a meromorphic map which extends to a meromorphic map $\tilde{f}: \overline{M} \longrightarrow N$. Take $p \in \mathbb{Z}[0,m]$. Then $A_p(\infty) < \infty$. (Griffiths-King [33], Proposition 5.9 and Carlson-Griffiths [8] Proposition 6.20).

<u>Proof.</u> Because $\tilde{f}$ is meromorphic and $\tilde{\omega}$ is real analytic, $f^*(c(L,\kappa))^p \wedge \tilde{\omega}^{m-1-p}$ is integrable over $K = \overline{M} - M(1)$ with

$$A_p(\infty) = A_p(1) + \int_K f^*(c(L,\kappa)^p \wedge \tilde{\omega}^{m-1-p} < \infty.$$

q.e.d.

The inverse is also true under stricter conditions. The following additional <u>General Assumptions</u> are made.

(E6) Let N be a connected, compact, complex manifold of dimension $n > 0$.

(E7) Let L be a holomorphic line bundle on N with a hermitian metric κ along the fibers of L such that $c(L,\kappa) > 0$.

(E8) Let $f: M \longrightarrow N$ be a holomorphic map.

By (E7) N is a projection algebraic variety. The map f is rational, if and only if f extends to a meromorphic map of $\overline{M}$ into N.

<u>Theorem 20.6.</u> Assume (E1) - (E8). Suppose $A(\infty) < \infty$. Then f is rational.

(Griffiths-King [33], Proposition 5.9 and Carlson-Griffiths [8], Proposition 6.20. If $M = \mathbb{C}^m$ and $N = \mathbb{P}_n$, the theorem was already proved in [67] Satz 24.1.)

Proof. By Kodaira [42] (see also Wells [88]) there exists $p \in \mathbb{N}$ such that $Q = L^p$ is ample and such that the dual classification map $\varphi: N \longrightarrow \mathbb{P}(V^*)$ with $V = \Gamma(N,Q)$ is smooth and injective. The evaluation map $\eta: N \times V \longrightarrow Q$ is an amplification of Q. Take a hermitian metric ℓ on V. A constant $\delta > 0$ exists such that $\tilde{\kappa} = \delta\, \kappa^p$ is a distinguished hermitian metric along the fibers of Q. Then $c(Q,\tilde{\kappa}) = p\, c(L,\kappa)$. Hence

$$A(\infty,Q,\tilde{\kappa},f) = p\ A(\infty,L,\kappa,f) = p\ A(\infty) < \infty.$$

Take $a \in \mathbb{P}(V)$. If $f^{-1}(E_Q[a]) = M$, its closure is analytic in $\overline{M}$. Assume $f^{-1}(E_Q[a]) \neq M$. Then f is adapted to a for Q. Take $0 < s < r$. Since $m_a(r,Q,\tilde{\kappa},f) \geqq 0$, the First Main Theorem implies

$$\frac{N_a(r,s,Q,f)}{\log r} \leqq \frac{T(r,s,Q,\tilde{\kappa},f)}{\log r} + \frac{m_a(s,Q,\tilde{\kappa},f)}{\log r}$$

Now, $r \longrightarrow \infty$ implies $n_a(\infty,Q,f) \leqq A(\infty,Q,\tilde{\kappa},f) < \infty$. Therefore $\theta_f^a[Q]$ is affine algebraic in W by Lemma 20.3. Hence the closure of $f^{-1}(E_Q[a]) = (\varphi\circ f)^{-1}(\ddot{E}[a])$ is analytic in $\overline{M}$ for each $a \in \mathbb{P}(V)$. By Proposition 20.1, $\varphi\circ f$ is meromorphic on $\overline{M}$. By Proposition 20.2, f extends to a meromorphic map on $\overline{M}$, hence f is rational; q.e.d.

Theorem 20.7. Assume (E1) - (E8). Suppose that M is smooth and that f has rank n. Take $w \in \mathbb{N}$. Define $p = [(K_N^*)^w/L]/w$. Let V be a linear subspace of $\Gamma(N,L)$ with $\dim V > 0$. Let $\eta: N \times V \longrightarrow L$ be the evaluation map. Let ℓ be a hermitian metric on V. Assume that κ is distinguished in respect to η and ℓ. Let $a_1,\ldots,a_q$ be given in $\mathbb{P}(V)$ such that $q > p$ and such that $a_1,\ldots,a_q$ are in general position for L. Let ν be the branching divisor of β. Take $\varepsilon > 0$ and $s > 0$. Then $T(r,s) \longrightarrow \infty$ for $r \longrightarrow \infty$ and a constant $c > 0$ exists such that

$$\|_{\varepsilon}\quad (q-p)\, T(r,s) \leqq \sum_{j=1}^{q} N_{a_j}(r,s) + N_\nu(r,s) + 2\varepsilon\, \varsigma\, n \log r$$

(20.5)

$$+ c \log T(r,s).$$

If f is rational, then

$$(20.6) \qquad \sum_{j=1}^{q} \delta_f(a_j,L) \leqq [K_N^*:L] + \frac{n_\nu(\infty)}{A(\infty)} < \infty.$$

If f is transcendental (= not rational), then

$$(20.7) \qquad \sum_{j=1}^{q} \delta_f(a_j,L) \leqq [K_N^*:L].$$

(Observe that (20.6) and (20.7) are trivial if $q \leqq [K_N^*:L]$.)

Proof. Theorem 18.14E applies and shows (20.5) with $T(r,s) \longrightarrow \infty$ for $r \longrightarrow \infty$. By Theorem 20.4, $n_\nu(\infty) < \infty$. Hence (20.5) implies (20.7), if f is transcendental. (Also Theorem 19.3 could be used). If f is rational, then $0 < A(\infty) < \infty$ and

$$\limsup_{r\to\infty} \frac{N_\nu(r,s)}{T_f(r,s)} = \limsup_{r\to\infty} \frac{N_\nu(r,s)}{\log r} \cdot \frac{\log r}{T(r,s)} = \frac{n_\nu(\infty)}{A(\infty)}.$$

Therefore (20.5) implies (20.6); q.e.d.

If $M = Y$ and if $\gamma: M \longrightarrow Y$ is the identity then $n_\nu(\infty) = 0$ and (20.7) also holds. In the case of (20.7), $f(M) \cap E_L[a_j] \neq \emptyset$ for at least one $j \in \mathbb{N}[1,q]$, which is the Picard-Borel Theorem.

(20.5) and (20.6) where proved by Griffiths-King [33]. In fact they derive the Defect-Relation in the algebraic case only. Also compare the results of this section with Griffiths-King [33] 2,4,5 page 195 and Carlson-Griffiths [8],5,6.

21. Applications and Variations

In sections 8 and 9 of [33] Griffiths and King give some applications and variations of their results. Three of these topics are selected for presentation and improvement here.

a) Holomorphic maps into algebraic varieties of general type.

Let N be a connected, projective algebraic manifold of dimension n. Let L be a holomorphic line bundle on N. According to Kodaira [43] and Kobayashi-Ochiai [40]

$$D(L) = \limsup_{k\to\infty} k^{-n} \dim \Gamma(N,L^k) < \infty.$$

Let K_N be the canonical bundle of N. According to Kodaira [43] N is said to be of <u>general type</u> if $D(K_N) > 0$. If $K_N > 0$, then N is of general type.

<u>Theorem 21.1.</u> Let N be a connected, n-dimensional, projective algebraic manifold of general type. Let L be a positive holomorphic line bundle on N with a hermitian metric κ along the fibers of L such that $c(L,\kappa) > 0$. Let M be a connected, parabolic complex manifold of dimension $m \geqq n$ with τ as parabolic exhaustion. Let $f: M \longrightarrow N$ be a holomorphic map. Let F be an effective Jacobian section of f dominated by τ with Y as dominator. Take $\theta > 0$. Then there exist positive constants c_1, c_2, c_3 such that for each $\varepsilon \in \mathbb{R}(0,\theta)$

$$(21.1) \qquad \|_{\varepsilon} \quad T(r,s) \leqq c_1 \log Y(r) + c_2 \mathrm{Ric}_{\tau}(r,s) + c_3 \varepsilon \log r.$$

Also $T(r,s) \longrightarrow \infty$ for $r \longrightarrow \infty$. The constants c_1, c_2 and c_3 do not depend on f and $\varepsilon \in \mathbb{R}(0,\theta)$.

<u>Proof.</u> An integer $p \in \mathbb{N}$ exists such that L^p is ample and such that the dual classification map $\varphi: N \longrightarrow \mathbb{P}(\Gamma(N,L^p))$ is smooth and injective. By Kobayashi-Ochiai [40] and Kodaira [43] $k \in \mathbb{N}$ exists such that $V = \Gamma(N,I)$ has positive dimension with $I = K^k \otimes (L^p)^*$. Let $\eta: N \times V \longrightarrow I$ be the evaluation map. Since rank $f = n$, the map f is adapted to each $a \in \mathbb{P}(V)$ for I. Also f is safe for I.

Let $\hat{\kappa}$ be a distinguished hermitian metric along the fibers of I. A form $\Omega > 0$ of class C^{∞} and degree 2n exists such that $\hat{\kappa} = (\kappa_{\Omega})^k \otimes (\kappa^*)^p$. The First Main Theorem holds

$$T(r,s,I,\hat{\kappa},f) = N_a(r,s,I,f) + m_a(r,I,\hat{\kappa},f) - m_a(s,I,\hat{\kappa},f)$$

for $0 < s < r$ and $a \in \mathbb{P}(V)$. Now $c(L,\hat{\kappa}) = k \,\mathrm{Ric}\, \Omega - p\, c(L,\kappa)$ implies

$$(21.2) \qquad \begin{aligned} &k \,\mathrm{Ric}\, (r,s,\Omega,f) - m_a(r,I,\hat{\kappa},f) + m_a(s,I,\hat{\kappa},f) \\ &= N_a(r,s,I,f) + p\, T(r,s,L,\kappa,f). \end{aligned}$$

Define $\xi_a \geqq 0$ by $\xi_a(x) = (\|a,x\|^{\wedge}_{\kappa})^{\frac{2}{k}}\Omega(x)$ for $x \in N$. Then $\xi_a > 0$ is of class C^∞ on $N - E_I[a]$ with k Ric $\xi_a = p\, c(L,\kappa) > 0$. A function $v \geqq 0$ of class C^∞ exists on $M - Z(F)$ such that $\upsilon^m = v\, F[\Omega]$. If $0 < s < r \in \mathfrak{E}^0_\tau$ and $s \in \mathfrak{E}^0_\tau$, then Theorem 15.5 implies

$$\text{Ric}_\tau(r,s) = N_{\theta_F}(r,s) + \text{Ric}\,(r,s,\Omega,f)$$

(21.3)

$$+ \tfrac{1}{2}\int_{M\langle r\rangle} \log v\,\sigma - \tfrac{1}{2}\int_{M\langle s\rangle} \log v\,\sigma.$$

Define $\zeta_a = (\|a,f\|_\kappa)^{\frac{2}{k}} v^{-1} \geqq 0$ on $M^+ - Z(F)$. Then

$$(21.4)\quad \mu_a(r) = \tfrac{1}{2}\int_{M\langle r\rangle} \log \zeta_a\sigma = -\tfrac{1}{2}\int_{M\langle r\rangle} \log v\,\sigma - \frac{1}{k}\, m_a(r,I,\hat{\kappa},f)$$

exists for $r \in \mathfrak{E}^0_\tau$. If $0 < s < r \in \mathfrak{E}^0_\tau$ with $s \in \mathfrak{E}^0_\tau$, this implies

$$\text{Ric}_\tau(r,s) + \mu_a(r) - \mu_a(s) =$$

$$= N_{\theta_F}(r,s) + \text{Ric}\,(r,s,\Omega,f) - \frac{1}{k}\, m_a(r,I,\hat{\kappa},f) + \frac{1}{k}\, m_a(s,I,\hat{\kappa},f)$$

$$= N_{\theta_F}(r,s) + \frac{1}{k}\, N_a(r,s,I,f) + \frac{p}{k}\, T(r,s,L,\kappa,f).$$

Hence

$$(21.5)\qquad T(r,s,L.\kappa,f) \leqq \frac{k}{p}\,(\text{Ric}_\tau(r,s) + \mu_a(r) - \mu_a(s))$$

Observe $F[\xi_a] = (\|a,f\|^{\wedge}_{\kappa})^{\frac{2}{k}}\, F[\Omega] = \zeta_a \upsilon^m$. Because ξ_a is continuous and $c(L,\kappa) > 0$, a constant $\gamma_1 > 0$ exists such that $(\gamma_1 c(L,\kappa))^n \geqq \xi_a$. On $\overset{+}{M}(r)$ this implies

$$\gamma_1 Y(r)\; f^*(c(L,\kappa)) \wedge \upsilon^{m-1} \geqq n\left(\frac{F[(\gamma_a c(L,\kappa))^n]}{\upsilon^m}\right)^{\frac{1}{n}} \upsilon^m$$

(21.6)

$$\geqq n\left(\frac{F[\xi_a]}{\upsilon^m}\right)^{\frac{1}{n}} \upsilon^m = n(\zeta_a)^{\frac{1}{n}}\, \upsilon^m.$$

If $t > 0$ and $0 < s < r$, define

$$(21.7) \qquad 0 < Q_a(t) = n\, t^{2-2m} \int_{M[t]} (\zeta_a)^{\frac{1}{n}} \upsilon^m < + \infty$$

$$(21.8) \qquad 0 < S_a(r,s) = \int_s^r Q_a(t) \frac{dt}{t} < \infty$$

because τ is strong. Now (21.6) - (21.8) imply

$$0 < S_a(r,s) \leqq \gamma_1 Y(r)\; T(r,s,L,\kappa,f).$$

Hence $T(r,s,L.\kappa,f) \longrightarrow \infty$ for $r \longrightarrow \infty$. The same proof as to Lemma 18.11 implies

$$2\mu_a(r) \leqq (2+\varepsilon)\, \varsigma\, n \log T(r,s,L,\kappa,f) + (2+\varepsilon)\, \varsigma\, n \log Y(r)$$

$$+ 4\, \varepsilon\, \varsigma\, n \log r + (2+\varepsilon)\, \varsigma\, n \log \gamma_1$$

for all $\varepsilon > 0$. Take $\varepsilon \in \mathbb{R}(0,\theta)$. Then (21.5) implies

$$T(r,s,L,k,f) \leqq \frac{k(2+\theta)\varsigma n}{2p} \log T(r,s,L,\kappa,f) + \frac{k}{p} \operatorname{Ric}_\tau(r,s)$$

$$(21.9)\,\|_\varepsilon \qquad + \frac{(2+\theta)\varsigma kn}{2p} \log Y(r) + \frac{2\varepsilon \varsigma nk}{p} \log r$$

$$+ \frac{k(2+\theta)\varsigma n}{2p} \log \gamma_1 - \frac{k}{p}\, \mu_a(s).$$

Define $c_1 = (2+\theta)\varsigma n\, k/p$ and $c_2 = 2k/p$ and $c_3 = 4\varsigma\, kn/p$. A constant $r_0 > s$ exists such that

(21.10)

$$\frac{c_1}{2} \log T(r,s,L,\kappa,f) + \frac{c_1}{2} \log \gamma_1 - {}^k/p\; \mu_a(s) \leqq \tfrac{1}{2}\, T(r,s,L,\kappa,f).$$

for all $r > r_0$. Hence (21.9) and 21.10) imply (21.1);q.e.d.

This estimate imposes restrictions on the growth of maps into algebraic manifolds of general type.

Corollary 21.2. Let N be a connected, n-dimensional projective algebraic manifold of general type. Let $f\colon \mathbb{C}^m \longrightarrow N$ be a holomorphic map. Then rank $f < n$ (Griffiths-King [33], Corollary 8.7).

Proof. If $m < n$, this is trivial. Assume $m \geqq n = \operatorname{rank} f$. Define τ by $\tau(\mathfrak{z}) = |\mathfrak{z}|^2$. Then $\operatorname{Ric}_\tau(r,s) = 0$. By Proposition 18.6 there exists an effective Jacobian section F of f dominated by τ with dominator $Y \equiv m$. Let L be a positive holomorphic line bundle on N with a hermitian metric κ along the fibers of L such that $c(L,\kappa) > 0$. Take $\varepsilon \in \mathbb{R}(0,1)$. Then $T(r,s) \longrightarrow \infty$ for $r \longrightarrow \infty$ and

$$\|_\varepsilon \qquad \frac{T(r,s)}{\log r} \leqq c_1 \frac{\log m}{\log r} + c_3\varepsilon .$$

Hence $r \longrightarrow \infty$ implies $0 \leqq A(\infty) \leqq c_3\varepsilon$. Now $\varepsilon \longrightarrow 0$ gives $A(\infty) = 0$. Hence $A(r) = 0$ and $T(r,s) = 0$, which contradicts $T(r,s) \longrightarrow \infty$ for $r \longrightarrow \infty$; q.e.d.

Corollary 21.3. Let M be a connected, n-dimensional projective algebraic manifold of general type. Let M be a connected affine algebraic manifold of dimension m. Let $f\colon M \longrightarrow N$ be a holomorphic map of rank n. Then f is rational. (Compare Kodaira [43], Griffiths-King [33], Proposition 8.1).

Proof. (E1) - (E8) can be satisfied. By Proposition 18.6 there exists an effective Jacobian section F of f dominated by τ with dominator $Y = m$. Then Theorem 21.1 holds with $\operatorname{Ric}_\tau(r,s) = O(\log r)$. Hence a constant $c > 0$ exists such that $\|_\varepsilon \; T(r,s) \leqq c \log r$. Therefore $A(\infty) \leqq c < \infty$. By Theorem 20.6 f is rational; q.e.d.

Let D be an unbounded subset of $\mathbb{R}_+$. The order of a function $g\colon D \longrightarrow \mathbb{R}$ is defined by

$$(21.11) \qquad \operatorname*{Ord}_D g = \limsup_{D \ni r \to \infty} \frac{\log^+ g(r)}{\log r} .$$

If $\mathbb{R}_+ - D$ is bounded $\operatorname*{Ord}_D g$ does not depend on D and is denoted by Ord g.

Let M be a parabolic manifold with a parabolic exhaustion τ of M. Then the order of $\operatorname{Ric}_\tau(r,s)$ on $\mathfrak{G}^0_\tau$ is called the order of τ and denoted by Ord τ. Obviously Ord τ does not depend on s. If ν is an analytic precycle on M define $\operatorname{Ord} \nu = \operatorname{Ord} n_\nu = \operatorname{Ord} N_\nu$. Let N be a compact, connected complex space and let L be a non-negative holomorphic line bundle on N with a hermitian metric κ along the fibers of L such that $c(L,\kappa) \geqq 0$. Let $f: M \longrightarrow N$ be a meromorphic map. Then the <u>order</u> of f for (L,κ) is defined by

$$\mathrm{Ord}(f,L,\kappa) = \mathrm{Ord}\, T(r,s,L,\kappa,f).$$

Obviously Ord (f,L,κ) does not depend on s. If $\chi > 0$ is a continuous form of bidegree (1.1) on N, define

$$T(r,s,\chi,f) = \int_s^r \int_{M[t]} f^*(\chi) \wedge \upsilon^{m-1} t^{1-2m} dt$$

$$\mathrm{Ord}\,(\chi,f) = \mathrm{Ord}\, T(\square,s,\chi,f).$$

Obviously, Ord (χ,f) does not depend on s.

<u>Lemma 21.4.</u> Ord (χ,f) does not depend on χ.

<u>Proof.</u> Let $\tilde{\chi} > 0$ be another continuous form of bidegree (1.1) on the compact space N. Then constants $\gamma_0 > 0$ and $\gamma_1 > 0$ exist such that $\gamma_0\tilde{\chi} \leqq \chi \leqq \gamma_1\chi$. Therefore

$$\gamma_0 T(r,s,\tilde{\chi},f) \leqq T(r,s,\chi,f) \leqq \gamma_1 T(r,s,\tilde{\chi},f).$$

Hence Ord (χ,f) = Ord $(\tilde{\chi},f)$;q.e.d.

Define Ord f = Ord (χ,f) as the <u>order</u> of f. If $c(L,\kappa) > 0$ then Ord (f,L,κ) = Ord f.

<u>Corollary 21.5.</u> Let N be a connected, n-dimensional, projective algebraic manifold of general type. Let M be a connected, parabolic manifold with parabolic exhaustion τ. Let $f: M \longrightarrow N$ be a holomorphic map with rank f = n = dim N = dim M. Then Ord f $\leqq$ Ord τ.

<u>Proof.</u> The effective Jacobian section F = Df of f is dominated by τ with a dominator $Y \equiv n$. Let L be a positive line bundle with a hermitian metric κ such that $c(L,\kappa) > 0$. Define ρ = Ord τ. If $\rho = \infty$, the statement is trivial. Assume $\rho < \infty$. Take $\varepsilon > 0$. A constant $r_0 > s$ exists such that

$$\mathrm{Ric}_\tau(r,s) \leqq r^{\rho+\frac{\varepsilon}{2}} \quad \text{for all } r \geqq r_0.$$

By Theorem 21.1

$$\|_{\rho+\varepsilon} \qquad T(r,s) \leqq r^{\rho+\varepsilon}.$$

Then $E = \{r \in \mathbb{R}(s,\infty) \mid T(r,s) > r^{\rho+\varepsilon}\}$ is open with

$$\int_E t^{\rho+\varepsilon} dt = C < \infty.$$

Take $x \in \mathbb{R}(s,\infty)$. If $x \in \mathbb{R}_+ - E$, then $T(r,s) \leqq r^{\rho+\varepsilon}$ if $x \in E$, then $y > x$ exists such that $\mathbb{R}[x,y) \subset E$ but $y \in \mathbb{R}_+ - E$. Therefore

$$T(r,s) \leqq T(y,s) \leqq y^{\rho+\varepsilon} = x^{\rho+\varepsilon} + (\rho+\varepsilon)\int_x^y t^{\rho+\varepsilon-1} dt$$
$$\leqq x^{\rho+\varepsilon} + \frac{\rho+\varepsilon}{x} C$$

Therefore $r_1 > s$ exists such that $T(r,s) \leqq 2\, r^{\rho+\varepsilon}$ for all $r \geqq r_1$. Hence $\mathrm{Ord}\, f = \mathrm{Ord}\, T(r,s) \leqq \rho$; q.e.d.

Is Corollary 21.5 sharp? Also examples of parabolic exhaustions of finite order would be welcome. Compare Griffiths-King [33]8a).

b) Variations of the Picard-Borel Theorem

Let N be a complex manifold of pure dimension n. Let ν be a divisor on N. Then ν is said to have normal crossings if and only if there exist divisors $\nu_1,..,\nu_q$ such that $\nu = \nu_1+\ldots+\nu_q$ and such that $\nu_1,..,\nu_q$ are in general position. Observe that $\nu_1,..,\nu_q$ may not be unique. Of course $\nu,\nu_1,..,\nu_q$ are holomorphic divisors.

For section b) the following assumptions will be made: Let N be a connected, projective algebraic manifold of dimension $n > 0$. Let L be a positive holomorphic line bundle on N with a hermitian metric κ along the fibers of L such that $c(L,\kappa) > 0$. Let K_N be the canonical bundle of N. Assume $L \otimes K_N$ is positive. Let ν be a divisor of a holomorphic section of L over N. Assume ν has normal crossings. Let M be a connected, parabolic complex manifold of dimension $m \geqq n$ with a parabolic exhaustion τ. Let $f\colon M \longrightarrow N$ be a holomorphic map.

Theorem 21.6. Let F be an effective Jacobion section of f dominated by τ with dominator Y. Then ν pulls back to a divisor ν_f on M. A number λ with $0 < \lambda < 1$ exists such that

$$(1-\lambda)\, T(r,s) \leqq 2\, N_{\nu_f}(r,s) + 2\, \mathrm{Ric}_\tau(r,s) +$$

$$(21.12) \quad \|_\varepsilon \qquad + (2+\varepsilon) \varsigma\, n \log Y(r) + 4\, \varepsilon\, n \varsigma \log r$$

for each $\varepsilon > 0$. Also $T(r,s) \longrightarrow \infty$ for $r \longrightarrow \infty$.

Proof. Holomorphic divisors $\nu_1, \ldots, \nu_q$ in general position exist such that $\nu = \nu_1 + \ldots + \nu_q$. For each $j \in \mathbb{N}[1,q]$ a holomorphic line bundle L_j and a section $0 \not\equiv \mathfrak{a}_j \in \Gamma(N, L_j) = V_j$ exist such $\nu_j = \theta_{\mathfrak{a}_j}$ and such that $L = L_1 \otimes \ldots \otimes L_q$. Define $a_j = \mathbb{P}(\mathfrak{a}_j) \in \mathbb{P}(V_j)$. Define $V = V_1 \otimes \ldots \otimes V_q$. Let $\eta_j \colon N \times V_j \longrightarrow L_j$ and $\eta \colon N \times V \longrightarrow L$ be the evaluation maps. Then ν is the divisor of $a = a_1 \otimes \ldots \otimes a_q$ and ν_j of a_j. Because f has rank n, f is adapted to a and a_j with $f^*(\nu_j) = \theta_f^{a_j}[L_j]$ and $\nu_f = f^*(\nu) = \theta_f^a[L]$. Hence

$$N_{\nu_f}(r,s) = N_a(r,s) = \sum_{j=1}^{q} N_{a_j}(r,s,L_j,f).$$

Hermitian metrics κ_j along the fibers of L_j exist such that $\kappa = \kappa_1 \otimes \ldots \otimes \kappa_q$. Let ℓ_j be a hermitian metric on V_j and define $\ell = \ell_1 \otimes \ldots \otimes \ell_q$ chosen such that κ is distinguished for η and ℓ. (C1) - (C5) and (C8) - (C12) are satisfied. On N a form $\Omega > 0$ of class C^∞ and degree 2n exists such that $c(L,\kappa) + \mathrm{Ric}\,\Omega > 0$. A rational number $\rho \in \mathbb{R}(0,1)$ exists such that $\rho\, c(L,\kappa) + \mathrm{Ric}\,\Omega > 0$ on N. Hence integers $0 < u < w$ with $w\rho = u$ exist such that $L^u \otimes K_N^w$ is positive. Therefore $0 \leqq p = [(K_N^*)^w/L] \leqq u < w$. Theorem 18.13 applies. Define $\lambda = {}^p/w \in \mathbb{R}(0,1)$. Take $\varepsilon > 0$. Then

$$(1-\lambda)\, T(r,s) \leqq N_{\nu_f}(r,s) + c_1 \log T(r,s) + \mathrm{Ric}_\tau(r,s)$$

$$(21.13) \quad \|_\varepsilon \qquad + c_2 \log Y(r) + c_3 \log r$$

where c_2 and c_3 are defined by (18.17). Also $T(r,s) \longrightarrow \infty$ for $r \longrightarrow \infty$.

A constant $r_0 > s$ exists such that $2c_1 \log T(r,s) \leqq (1-\lambda)\, T(r,s)$ for all $r > r_0$. Hence (21.13) implies (21.12);q.e.d.

Corollary 21.7. Assume $M = \mathbb{C}^m, \tau(\mathfrak{z}) = |\mathfrak{z}|^2$ and $f(\mathbb{C}^m) \cap \operatorname{supp} \nu = \emptyset$. Then rank $f < n$. (Griffiths-King [33], Proposition 8.8 and Carlson-Griffiths [8], Corollary (6.11)).

Proof. If rank $f = n$, then there exists an effective Jacobian section F of f dominated by τ with dominator $Y = m$. Also $N_{\nu_f}(r,s) \equiv 0 \equiv \operatorname{Ric}_\tau(r,s)$. If f is transcendental, (21.12) implies $0 < 1 - \lambda \leqq 0$ which is impossible, if f is of rational growth (21.12) implies $0 < 1 - \lambda < 4\, n\, \varsigma\, \varepsilon/A(\infty)$ for all $\varepsilon > 0$ which is also impossible. Hence rank $f < n$;q.e.d.

Corollary 21.8. Assume M is a connected, affine algebraic manifold of dimension m embedded into a complex vector space W. Assume the pull back divisor ν_f of ν is affine algebraic. Assume rank $f = n$. Then f is rational. (Griffiths-King [33], Propositon 8.8 and Carlson-Griffiths [8], Corollary 6.21).

Proof. (E1) - (E8) can be realized. An effective Jacobian section F exists dominated by τ with a dominator $Y \equiv m$. Then $N_{\nu_f}(r,s) = O(\log r)$ and $\operatorname{Ric}_\tau(r,s) = O(\log r)$ by Lemma 20.3 and Theorem 20.4. Take $\varepsilon > 0$. Then $\lambda \in \mathbb{R}(0,1)$ and $\gamma_0 > 0$ exist such that $\|_\varepsilon (s-\lambda)\, T(r,s) \leqq \gamma_0 \log r$ by (21.12). Hence $A(\infty) < \infty$. Now Theorem 20.6 shows that f is rational.

Corollary 21.9. Assume $m = n =$ rank f. Then ν pulls back to a divisor ν_f on M with $\operatorname{Ord} f \leqq \operatorname{Max}(\operatorname{Ord} \tau, \operatorname{Ord} \nu_f)$.

Proof. Define $\rho = \operatorname{Max}(\operatorname{Ord} \tau, \operatorname{Ord} \nu_f)$. If $\rho = \infty$, the statement is trivial. Assume $\rho < \infty$. The Jacobian section $F = Df$ of f is effective and is dominated by τ with a dominator $Y = n$. Theorem 21.6 applies with some constant $\lambda \in \mathbb{R}(0,1)$. Take $\varepsilon > 0$. A constant $r_0 > s$ exists such that

$$2N_{\nu_f}(r,s) + 2 \operatorname{Ric}_\tau(r,s) + (2+\varepsilon)\, n\, \varsigma \log n + 4 \varsigma\, \varepsilon\, n \log r \leqq (1-\lambda) r^{\rho+\varepsilon}$$

for all $r \geqq r_0$. Therefore $\|_\varepsilon T(r,s) \leqq r^{\rho+\varepsilon}$. Hence $\operatorname{Ord} f \leqq \rho$ (see proof of Corollary 21.5);q.e.d.

Compare the results of this section b) with Griffiths-King [33], 36 and Carlson-Griffiths [8], Proposition 6.19.

c) Theorem on the logarithmic derivative

Griffiths-King [33] extend the well known Theorem on the logarithmic derivative to several variables. Their proof is based on an estimate of $\mu_a(r)$ if $\lambda > 1$. This shall be investigated first.

Assume (C1) - (C12) with $\lambda > 1$ and $c(L,\kappa) > 0$. Define μ_a by (17.1). For $r \in \mathfrak{C}^0_\tau$ the following integral exists

$$\mu_a^+(r) = \tfrac{1}{2}\int_{M\langle r\rangle} \log^+\zeta_a \ \sigma \geqq 0. \tag{21.14}$$

Clearly $\mu_a(r) \leqq \mu_a^+(r)$. Define

$$0 \leqq w = \prod_{j=1}^{q} (\delta u_j) = \delta^q \|a,\square\|_\kappa^2 < 1 \tag{21.15}$$

on N. Then $w > 0$ on $N - E_L[a]$. Observe $\zeta_a > 0$ on $Z_a - Z(F)$. Define

$$G = \frac{1}{n}\log^+\zeta_a + \frac{\lambda-1}{n}\log w\circ f \tag{21.16}$$

on $Z_a - Z(F)$. For $t > 0$ and $0 < s < r$, define

$$0 < Q_a^*(t) = n\, t^{2-2m}\int_{M[t]} e^G\, \upsilon^m \leqq +\infty. \tag{21.17}$$

$$0 < S_a^*(r,s) = \int_s^r Q_a^*(t)\,\frac{dt}{t} \leqq +\infty. \tag{21.18}$$

Lemma 21.10. Assume (C1) - (C12) with $\lambda > 1$. If $0 < s < r$, then

$$0 < S_a^r(r,s) \leqq Y(r)\ \mathrm{Ric}\,(r,s,\xi_a,f) + \frac{n\,\mathfrak{C}}{2}(r^2-s^2) < +\infty \tag{21.19}$$

$$\frac{\mathrm{Ric}(r,s,\xi_a,f)}{\log r} \longrightarrow \mathrm{ric}(\infty,\xi_a,f) > 0 \quad \text{for } r \longrightarrow \infty. \tag{21.20}$$

Proof. $(\mathrm{Ric}\ \xi_a)^n \geqq w^{\lambda-1}\xi_a$ by (16.8). Hence

$$Y(r)\ f^*(\mathrm{Ric}\ \xi_a)\wedge \upsilon^{m-1} \geqq n\left(\frac{F[(\mathrm{Ric}\ \xi_a)^n]}{\upsilon^m}\right)^{\frac{1}{n}} \upsilon^m$$
$$\geqq n(w\circ f)^{\frac{\lambda-1}{n}}\,(\zeta_a)^{\frac{1}{n}}\,\upsilon^m > 0$$

on $Z_a - Z(F)$. Hence $\mathrm{ric}(\infty,\xi_a,f) \geqq \mathrm{ric}(r,\xi_a,f) > 0$ and (21.20) follows.

If $x \geqq 0$ and $0 < y \leqq 1$, then $xy \geqq \exp(\log^+ x + \log y) - 1$. Hence

$$(w\circ f)^{\frac{\lambda-1}{n}} (\zeta_a)^{\frac{1}{n}} \geqq e^G - 1$$

and

$$Y(r)\ \mathrm{Ric}(r,s,\xi_a,f) \geqq \int_s^r t^{1-2m} \int_{M[t]} n(w\circ f)^{\frac{\lambda-1}{n}} (\zeta_a)^{\frac{1}{n}} \upsilon^m dt$$

$$\geqq S_a^*(r,s) - \int_s^r n t^{1-2m} \int_{M[t]} \upsilon^m dt$$

$$= S_a^*(r,s) - \frac{n\varsigma}{2}(r^2 - s^2)$$

q.e.d.

<u>Lemma 21.11.</u> Assume (C1) - (C17) with $\lambda > 1$. Then

(21.21) $$\hat{Q}_a^*(t) = \frac{1}{\varsigma} \int_{M\langle t\rangle} e^G \sigma > 0$$

exists for almost all $t > 0$. If $r > 0$, then

(21.22) $$r^{2m-2} Q_a^*(t) = 2mn\varsigma \int_0^r \hat{Q}_a^*(t)\, t^{2m-1} dt$$

For almost all $r > 0$

(21.23) $$\frac{d}{dr}[r^{2m-2} Q_a^*(r)] = 2nm\,\varsigma\, \hat{Q}_a^*(r) r^{2m-1}$$

(21.24) $$2\mu_a^+(r) \leqq \varsigma n \log \hat{Q}_a^*(r) + (\lambda-1)(2m_a(r) - q \log \delta).$$

<u>Proof.</u> Take $r > 0$. Then

$$r^{2m-2} Q_a^*(r) = n\, m \int_{M[r]} e^G \tau^{m-1} d\tau \wedge \sigma$$

$$= 2nm \int_0^r \int_{M\langle t\rangle} e^G\, \sigma\, t^{2m-1} dt.$$

Hence $\hat{Q}_a^*(t)$ exists for almost all $t > 0$ and (21.22) holds which implies (21.23). The convexity of the logarithm implies

$$\log \hat{Q}_a^*(r) \geqq \frac{1}{\varsigma} \int_{M\langle r\rangle} G\,\sigma = \frac{2}{\varsigma n}\, \mu_a^+(r) + \frac{\lambda-1}{\varsigma n} \int_{M\langle r\rangle} \log(\delta^q \|a,f\|_\kappa^2)\,\sigma$$

$$= \frac{2}{\varsigma n}\, \mu_a^+(r) - \frac{(\lambda-1)}{\varsigma n}\, 2m_a(r) + \frac{\lambda-1}{\varsigma n}\, q \log \delta.$$

q.e.d.

Lemma 21.12. Assume (C1) - (C12) with $\lambda > 1$. Take $\varepsilon > 0$ and $s > 0$. Then

$$2\mu_a^+(r) \leqq (2+\varepsilon)\varsigma\, n \log \mathrm{Ric}(r,s,\xi_a,f) + 2(\lambda-1)\, m_a(r)$$

(21.15) $\|_\varepsilon$

$$+ (2+\varepsilon)\varsigma\, n \log Y(r) + 2(2+3\varepsilon)\varsigma\, n \log r.$$

Proof. Take $\beta \in \mathbb{R}(0,\varepsilon)$ such that $\varepsilon(\beta+2) + \beta(2m-1) < 4\varepsilon$. Define $\nu = 1 + \beta/2$. The same proof as to Lemma 18.11 shows

$\|_\varepsilon$
$$Q_a^*(x) \leqq \frac{x^{3\varepsilon}}{(2mn\varsigma)^\nu}\, (S_a^*(x,s))^{2\nu}.$$

Therefore

$$2\mu_a^+(r) \leqq 2\varsigma n\nu \log S_a^*(r,s) + 3\varepsilon\varsigma n \log r$$

(21.25)

$$+ 2(\lambda-1)\, m_a(r) - (\lambda-1)\, q \log \delta - n\varsigma\nu \log 2mn\varsigma.$$

A number $r_0 > s + 1$ exists such that $\mathrm{Ric}(r,s,\xi_a,f) > 1$ for all $r > r_0$. Hence 21.19 implies

(21.27) $$\log S_a^*(s,r) \leqq \log Y(r) + \log \mathrm{Ric}(r,s,\xi_a,f) + 2 \log r + C_0$$

for some constant C_0 and all $r > r$. Now (21.26) and (21.27) imply (21.25); q.e.d.

Theorem 21.13. Assume (C1) - (C12) with $\lambda > 1$ and $c(L,\kappa) > 0$. Take $\varepsilon > 0$ and $s > 0$. Then $T(r,s) \longrightarrow \infty$ for $r \longrightarrow \infty$ and

$$(21.28)\|_\varepsilon \qquad \begin{aligned} 2\mu_a^+(r) &\leqq 2(\lambda-1)T(r,s) + (2+\varepsilon)\varsigma n \log T(r,s) \\ &+ (2+\varepsilon)\varsigma n \log Y(r) + 4n\varsigma(1+2\varepsilon) \log r. \end{aligned}$$

(Griffiths-King [33], 9.16).

Proof. A constant $\gamma_1 > 0$ exists such that

$$- \gamma_1 c(L,\kappa) \leqq \mathrm{Ric}\ \Omega \leqq \gamma_1 c(L,\kappa).$$

Therefore $|\mathrm{Ric}(r,s,\Omega,f)| \leqq \gamma_1 T(r,s)$. By Proposition 18.12 a constant C_1 exists such that

$$(21.24) \qquad \begin{aligned} \mathrm{Ric}(r,s,\xi_a,f) &\leqq \lambda\, T(r,s) + \mathrm{Ric}(r,s,\Omega,f) + C_1 \\ &\leqq (\lambda+\gamma_1)\ T(r,s) + C_1. \end{aligned}$$

Since Ric $(r,s,\xi_a,f) \longrightarrow \infty$ for $r \longrightarrow \infty$, also $T(r,s) \longrightarrow \infty$ for $r \longrightarrow \infty$. Also $m_a(r) \leqq T(r,s) + m_a(s)$ by the First Main Theorem. Hence (21.29) and (21.25) imply (21.28) easily; q.e.d.

Corollary 21.14. Assume (C1) - (C12) with $\lambda > 1$ and $c(L,\kappa) > 0$. Let f have transcendental growth. Then

$$\liminf_{r\to\infty} \frac{\mu_a^+(r)}{T(r,s)} \leqq \lambda - 1 + n\,\varsigma\, Y_F.$$

If f has rational growth, then

$$\liminf_{r\to\infty} \frac{\mu_a^+(r)}{T(r,s)} \leqq \lambda - 1 + n\,\varsigma\, Y_F + \frac{2n\varsigma}{A(\infty)}.$$

Theorem 21.15. Let N be a connected, projective algebraic manifold of dimension n with a negative canonical bundle K_N. Let κ be a hermitian metric along the fibers of K_N^* such that $c(K_N^*,\kappa) > 0$. Let ψ be a meromorphic form of degree n on N without zeroes whose pole divisor $\nu = \nu_\psi^\infty$

has normal crossings. Let q be the number of branches of the pole set $\operatorname{supp} \nu$. Let M be a connected complex manifold of dimension m with a parabolic exhaustion τ. Let $f: M \longrightarrow N$ be a holomorphic map. Let F be an effective Jacobian section of f dominated by τ with dominator Y. Let κ_0 be the hermitian metric along the fibers of $K_M|M^+$ induced by υ^m. Then

$$(21.30) \qquad D(r) = \int_{M\langle r\rangle} \log^+ |F[\psi]|_{\kappa_0}\, \sigma \geqq 0$$

exists for almost all $r > 0$. For $\lambda > 1$ and $\varepsilon > 0$

$$\begin{aligned} D(r) &\leqq (\lambda-1)\, T(r,s,K_N^*,\kappa,f) + (1+\varepsilon)\, \varsigma\, n \log Y(r) \\ (21.31)\|_\varepsilon \qquad &+ (q+1+\varepsilon)\, \varsigma\, n \log T(r,s,K_N^*,\kappa,f) \\ &+ 4\, n\, \varsigma\, (1+\varepsilon) \log r. \end{aligned}$$

Also $T(r,s,K_N^*,\kappa,f) \longrightarrow \infty$ for $r \longrightarrow \infty$. If f has transcendental growth, then

$$(21.32) \qquad 0 \leqq D_f = \liminf_{r\to\infty} \frac{D(r)}{T(r,s,K_N^*,\kappa,f)} \leqq \varsigma\, n\, Y_F$$

If f has rational growth, then

$$(21.33) \qquad 0 \leqq D_f \leqq \varsigma\, n\, Y_F + \frac{4n\,\varsigma}{A(\infty,K_N^*,\kappa,f)}.$$

<u>Remark.</u> If $m = n$ and $F = Df$, then $Y_F = 0$. Therefore $D_f = 0$ in this case. (Griffiths-King [33], Proposition 9.3 with $M = \mathbb{C}^m$.)

<u>Proof.</u> Abbreviate $L = K_N^*$ and $T(r,s) = T(r,s,L,\kappa,f)$ etc. Divisors $\nu_1,\ldots,\nu_q$ in general position exist such that $\nu = \nu_1+\ldots+\nu_q$. Let $L_1,\ldots,L_q$ be the holomorphic line bundles defined by $\nu_1,\ldots,\nu_q$ with $L = L_1\otimes\ldots\otimes L_q$. Define $V_j = \Gamma(N,L_j)$ and $V = V_1\otimes\ldots\otimes V_q$. Let $\eta_j: N \times V_j \longrightarrow L_j$ and $\eta: N \times V \longrightarrow L$ be the evaluation maps. Take $\mathfrak{a}_j \in V_j$ such that ν_j is the divisor of $\mathfrak{a}_j$. Then ν is the divisor of $\mathfrak{a} = \mathfrak{a}_1\otimes\ldots\otimes\mathfrak{a}_q \in V$. Hence $(\mathfrak{a},\psi)$ is a nowhere vanishing holomorphic function on N. Hence $(\mathfrak{a},\psi) = 1$ can be assumed. Let ℓ_j be a hermitian metric on V with $|\mathfrak{a}_j| = 1$. Then $\ell = \ell_1\otimes\ldots\otimes\ell_q$ is a hermitian metric on V with $|\mathfrak{a}| = 1$. Let κ_j be a hermitian metric on L_j distinguished

for η_j and ℓ_j such that $\theta\kappa = \kappa_1 \otimes \ldots \otimes \kappa_q$ where θ is a constant. Since $c(L,\theta\kappa) = c(L,\kappa)$, w.$\ell$.o.g. $\theta = 1$ can be assumed. Then κ is distinguished for η and ℓ. Define $a_j = \mathbb{P}((\mathfrak{a}_j) \in \mathbb{P}(V_j)$ and $a = a_1 \otimes \ldots \otimes a_q = \mathbb{P}(\mathfrak{a}) \in \mathbb{P}(V)$. Then $a_1, \ldots, a_q$ are in general position. Define $u_j = \|\square, a_j\|^2_{\kappa_j}$. A form $\Omega > 0$ of class C^∞ and of degree 2n exists on N such that $\kappa_\Omega = \kappa^*$. Hence $c(L,\kappa\ ) + \text{Ric}\ \Omega = 0$. Hence $\lambda\ c(L,\kappa) + \text{Ric}\ \Omega > 0$ for each $\lambda > 1$. By Theorem 16.3 a Carlson-Griffiths form ξ_a exists satisfying (16.5) - (16.9). Hence (C1) - (C12) are satisfied for each fixed $\lambda > 1$.

Because $(\mathfrak{a}, \psi) = 1$ also $|\mathfrak{a}|^2_\kappa = |\psi|^{-2}_{\kappa^*}$ on $N - \text{supp}\ \nu$.
Also $u_1 \ldots u_q = |\mathfrak{a}|^2_\kappa$ and $i_n \psi \wedge \overline{\psi} = |\psi|^2_{\kappa^*}\ \Omega$ by (2.2). Hence

$$\xi_a = \gamma\ \delta^{-\lambda q}(|\mathfrak{a}|^2_\kappa)^{1-\lambda} \prod_{j=1}^{q} (\log \delta u_j)^{-2} i_n \psi \wedge \overline{\psi}.$$

Define $\Delta = |F[\psi]|_{\kappa_0}$. Again (2.2) implies

$$F[i_n \psi \wedge \overline{\psi}] = i_m F[\psi] \wedge \overline{F[\psi]} = \Delta^2\ \upsilon^m.$$

Also $|\mathfrak{a}|_\kappa \circ f = \|a, f\|_\kappa$. Hence

$$F[\xi_a] = \gamma\ \delta^{-\lambda q} \|a, f\|^{2(1-\lambda)}_\kappa \prod_{j=1}^{q} (\log \delta u_j \circ f)^{-2}\ \Delta^2\ \upsilon^m.$$

Also $F[\xi_a] = \zeta_a \upsilon^m$. Hence

$$\Delta^2 = \frac{\delta^{\lambda q}}{\gamma} \|a, f\|^{2(\lambda-1)}_\kappa \left(\prod_{j=1}^{q} \log \delta u_j \circ f\right)^2 \zeta_a.$$

Also $(\log \delta u_j)^2 \geqq 1$ and $\delta^q \|a, f\|^2_\kappa \leqq 1$. Therefore

$$2 \log^+ \Delta \leqq \log^+ \zeta_a + \sum_{j=1}^{q} \log\ (\log \delta u_j \circ f)^2 + q\ |\ \log \delta\ | + |\ \log \gamma|$$

Also

$$\sum_{j=1}^{q} \log\ (\log \delta u_j \circ f)^2 \leqq 2\ q\ \log\ \left(\frac{1}{q} \sum_{j=1}^{q} \log \frac{1}{u_j \circ f} + \log \frac{1}{\delta}\right)$$

$$= 2\ q\ \log\ \left(\frac{2}{q} \log \frac{1}{\|a, f\|_\kappa} + \log \frac{1}{\delta}\right).$$

Hence

$$\int_{M\langle r\rangle} \log\,(\log \delta u_j \circ f)^2\,\sigma$$

$$\leqq 2\,q\,\varsigma\,\log\,\Big(\frac{2}{q\,\varsigma}\int_{M\langle r\rangle} \log \frac{1}{\|a,f\|_\kappa}\,\sigma + \log\frac{1}{\delta}\Big)$$

$$\leqq 2\,q\,\varsigma\,\log^+ T(r,s) + C_1$$

for $0 < s < r$ where C_1 is some constant. Therefore

$$D(r) \leqq \mu_a^+(r) + q\,\varsigma\,\log^+ T(r,s) + C_2$$

for almost all $r > s > 0$, where C_2 is some constant. Hence

$$\|_\varepsilon \qquad D(r) \leqq (\lambda - 1)\,T(r,s) + \varsigma\,n(q+1+\varepsilon)\,\log^+ T(r,s)$$

$$+ (1+\varepsilon)\,\varsigma\,n\,\log Y(r) + 4n\,\varsigma(1+\varepsilon)\,\log r$$

which proves (21.31). By Theorem 21.13. $T(r,s) \longrightarrow \infty$ for $r \longrightarrow \infty$. Hence (21.31) implies (21.32) respectively (21.33) immediately, q.e.d.

Consider the case where $m = n$, and $M = \mathbb{C}^n$ and $\tau(\mathfrak{z}) = |\mathfrak{z}|^2$ and $F = Df$. Then f has rank n and $Y \equiv n$. Hence $Y_F = 0$. The canonical bundle K_M is trivial and

$$F[\psi] = f^*(\psi) = h\,dz_1 \wedge \ldots \wedge dz_n.$$

The function $h \not\equiv 0$ is meromorphic on $\mathbb{C}^n$. Also

$$\Delta^2 \upsilon^n = i_n F[\psi] \wedge \overline{F[\psi]} = |h|^2 \upsilon^n$$

implies $\Delta = |h|$ and

$$D(r) = \int_{\mathbb{C}^n\langle r\rangle} \log^+ |h|\,\sigma$$

consider $h\colon \mathbb{C}^n \longrightarrow \mathbb{P}_1 = \mathbb{C} \cup \{\infty\}$ as a meromorphic map. Then

$$m_\infty(r,h) = \int_{\mathbb{C}^n\langle r\rangle} \log \sqrt{1+|h|^2}\,\sigma$$

is the compensation function of h with

$$D(r) \leqq m_\infty(r,h) \leqq D(r) + \tfrac{1}{2}\log 2$$

If $f\colon \mathbb{C}^n \longrightarrow N$ is transcendental, this implies

$$D_f = \liminf_{r\to\infty} \frac{m_\infty(r,h)}{T(r,s,K_N^*,\kappa,f)} = 0.$$

If $N = \mathbb{P}_n$, then one and only one meromorphic form ψ exists on $\mathbb{P}_n$ such that

$$\mathbb{P}^*(\psi) = \sum_{j=0}^{n} (-1)^j \bigwedge_{j\neq\lambda=0}^{n} \frac{dw_\lambda}{w_\lambda} .$$

The pole divisor of ψ is in general position and ψ has no zeros. An irreducible holomorphic representation $\mathfrak{v}\colon \mathbb{C}^n \longrightarrow \mathbb{C}^{n+1}$ of f exists. Then $\mathbb{P}\circ\mathfrak{v} = f$ and $\mathfrak{v} = (v_0,\dots v_n)$ imply

$$f^*(\psi) = \sum_{j=0}^{n} (-1)^j \bigwedge_{j\neq\lambda=0}^{n} \frac{dv_\lambda}{v_\lambda} .$$

Here $f_\lambda = v_\lambda/v_0$ is a meromorphic function on $\mathbb{C}^n$. Then

$$f^*(\psi) = \frac{df_1\wedge\dots\wedge df_n}{f_1\dots f_n} = \frac{f'}{f_1\dots f_n}\, dz_1\wedge\dots\wedge dz_n$$

where f' is the Jacobian of $f = (f_1,\dots,f_n)$. Hence $h = f'/(f_1\dots f_n)$. Here $\mathbb{P}_n$ carries the usual Fubini Study Kaehler metric which defines a hermitian metric κ along $K^*_{\mathbb{P}_n}$. Then

$$c(K^*_{\mathbb{P}_n},\kappa) = (n+1)\ddot{\omega}$$

Hence $T(r,s,K^*_{\mathbb{P}_n},\kappa,f) = T(r,s,f)$ where

$$T(r,s,f) = \int_s^r t^{1-2n} \int_{\mathbb{C}^n[t]} f^*(\ddot{\omega}) \wedge \upsilon^{n-1}\, dt$$

is the classical characteristic of the meromorphic map f. Therefore the following result is obtained.

Theorem 21.16. Let $f\colon \mathbb{C}^n \longrightarrow \mathbb{P}_n = \overline{\overline{\mathbb{C}}}^n$ be a transcendental meromorphic map. Write $f = (f_1,\ldots,f_n)$ with meromorphic functions $f_1,\ldots,f_n$. Let f' be then Jacobian determinant. Then

$$\liminf_{r\to\infty} \frac{m_\infty\left(r, \dfrac{f'}{f_1,\ldots,f_n}\right)}{T(r,s,f)} = 0$$

(Griffiths-King [33] Proposition 9.3).

Therefore Theorem 21.15 and Theorem 21.16 can be considered generalizations of the theorem on the logarithmic derivative in one variable. Compare the results of this section c) with Griffiths-King [33] 8 and 9 and with Carlson-Griffiths [8], 6.

Appendix

Some more technical details needed in the proofs of the Main Theorems are put into this appendix in order to ease the flow of the representation of the general theory. The proofs of this appendix are based on results of Tung [87]

a) The singular Stokes and Residue Theorems

A family $\{g_\varepsilon\}_{\varepsilon\in\mathbb{R}(0,1)}$ of functions of class C^∞ is said to be a test family if $0 \leqq g_\varepsilon \leqq 1$, if $g_\varepsilon(x) = 1$ for all $x \geqq \varepsilon$, if $g_\varepsilon(x) = 0$ for all $x \leqq \frac{\varepsilon}{2}$ and if a constant $C > 0$ exists such that $\varepsilon|g'_\varepsilon(x)| \leqq C$ for all $x \in \mathbb{R}$ and all $\varepsilon \in \mathbb{R}(0,1)$. Test families exist.

In this section M is a complex space of pure dimension m and $\tau\colon \mathbb{C}^n \longrightarrow \mathbb{R}$ is defined by $\tau(\mathfrak{z}) = |\mathfrak{z}|^2$. Define ω and σ by (1.1) and (1.2).

Theorem A1. Let $f\colon M \longrightarrow \mathbb{C}^n$ be a q-fibering, holomorphic map with $q = m - n \geqq 0$. Let G be open and let $\overline{G}$ be compact in M. Let χ be a continuous form of bidegree (q,q) on M. Define $T = \mathrm{supp}_{\partial G}\chi$. Assume that $T \cap f^{-1}(0)$ has measure zero on $F = f^{-1}(0)$. Let $\{g_\varepsilon\}$ be a test family. Define $h_\varepsilon = g_\varepsilon(|f|)$. Then the integral

$$\psi(\varepsilon) = \int_G dh_\varepsilon \wedge f^*(\sigma) \wedge \chi$$

exists for each $\varepsilon \in \mathbb{R}(0,1)$ and $\psi(\varepsilon) \longrightarrow \int_{G\cap F} \nu_f\chi$ for $\varepsilon \longrightarrow 0$.

For the proof see Tung [87] Theorem 5.2.15.

Lemma A2. Let $f\colon M \longrightarrow \mathbb{C}^k$ be a holomorphic map of pure rank $n > 0$. Let K be compact in M. Let u,t,s be non-negative integers with $t \leqq s < 2n$ and $t \leqq 2n - 2$. Let χ be a form of degree $2m - s - 1$ on M and let ζ be a form of degree s on $\mathbb{C}^k$. Assume that χ and ζ are measurable and locally bounded and that $f^*(\zeta)$ exists almost everywhere on M and is measurable. Let $\{g_\varepsilon\}$ be a test family. Define $h_\varepsilon = g_\varepsilon(|f|)$ for each $\varepsilon \in \mathbb{R}(0,1)$. Define

$$\psi(\varepsilon) = \int_K |dh_\varepsilon \wedge (\log |f|)^u |f|^{-t} f^*(\zeta) \wedge \chi|.$$

Then $\psi(\varepsilon) \to 0$ for $\varepsilon \longrightarrow 0$. (Tung [87], 6.26)

<u>Lemma A3.</u> Let $f: M \longrightarrow \mathbb{C}^k$ be a holomorphic map of pure rank n. Let s,u and v be non-negative integers with $0 \leqq v \leqq s < 2n$. Let Y be an open neighborhood of 0 in $\mathbb{C}^k$. For each $y \in Y$, let χ_y be a measurable form of degree $2m - s$ on M and let ζ_y be a measurable form of degree s on $\mathbb{C}^k$ such that $f^*(\zeta_y)$ exists and is measurable on M. Assume that χ_y and ζ_y are locally uniformily bounded. Let K be compact in M. Then

$$\psi(y) = \int_K \left(\log \frac{1}{|f-y|}\right)^u \frac{f^*(\zeta_y) \wedge \chi_y}{|f-y|^v}$$

exists for all $y \in Y$ and ψ is measurable and locally bounded . Define

$$W = \{(w,y) \in \mathbb{C}^k \times Y \mid w \neq y\} \qquad Z = \{(x,y) \in M \times Y \mid f(x) \neq y\}.$$

If in addition ζ_y is continuous on W and if χ_y is continuous on Z, then ψ is continuous on Y. (Tung [87], 6.2.8).

Let G be an open subset of M and let $j: dG \longrightarrow M$ be the inclusion map. Let F be a thin analytic subset of M. Let ξ be a continuous, real form of degree $2m - 1$ on $M - F$. Define

$$\Gamma_+ = \{ x \in dG - F \mid j^*(\xi)(x) > 0\}$$

$$\Gamma_- = \{ x \in dG - F \mid j^*(\xi)(x) < 0\}.$$

Then ξ is said to have <u>separate signs on dG</u> if $\overline{\Gamma}_+ \cap \overline{\Gamma}_- = \emptyset$.

Let $\psi: M \longrightarrow \mathbb{R}$ be a continuous function. Then ψ is said to have a <u>consistent sign around ∂G</u> if an open neighborhood W of ∂G exists such that either $\psi|G \cap W \geqq 0$ or $\psi|G \cap W \leqq 0$ or $\psi|(W-G) \leqq 0$ or $\psi|(W-G) \geqq 0$ holds and if $\psi|\partial G = 0$. If so and if $\chi \geqq 0$ is a non-negative form of bidegree $(m-1,m-1)$ on M, then $j^*(d^c\psi \wedge \chi) \geqq 0$ on dG if $\psi|G \cap W \leqq 0$ or $\psi|(W-G) \geqq 0$ and $j^*(d^c\psi \wedge \chi) \leqq 0$ if $\psi|G \cap W \geqq 0$ or $\psi|(W-G) \leqq 0$, (see Lemma 7.2.)

<u>Theorem A.4. (Singular Stokes Theorem).</u> Let Y be an open neighborhood of 0 in $\mathbb{C}^k$. Let $f: M \longrightarrow Y$ be a holomorphic map of pure rank n. Let s,t and u be non-negative integers with $t \leqq s \leqq 2n - 2$. Let χ be a form of class C^1 and degree $2m - s - 1$ on M. Let G be Stokes admissible for χ. Let ζ be a form of class C^1 and degree s on $Y - \{0\}$. Assume

that ζ and $\sqrt{\tau}\, d\zeta$ are locally bounded on Y. Define

$$\xi = (\log |f|)^u \, |f|^{-t} f^*(\zeta) \wedge \chi.$$

Assume that either ξ is integrable over dG or has separate signs on G. Then ξ is integrable over dG and $d\xi$ is integrable over G with

$$\int_G d\xi = \int_{dG} \xi. \tag{1}$$

For the proof, take a test family $\{g_\varepsilon\}$. Define $h_\varepsilon = g_\varepsilon(|f|)$ and apply Stokes Theorem to $h_\varepsilon \xi$. Then the terms of $d(h_\varepsilon \xi)$ are computed explicitly. Now Lemma A.2 and Lemma A.3 can be applied to obtain (1). For details see [82] Theorem 9.4.

Remark. Since $\tau\,\omega$ is bounded Theorem 9.4 applies to the forms

$$\xi = (\log |f|)^u f^*(\omega^s) \wedge \chi \qquad \text{if } 0 \leqq s \leqq n-1$$

$$\xi = (\log |f|)^u f^*(d^c \log \tau \wedge \omega^s) \wedge \chi \qquad \text{if } 0 \leqq s \leqq n-2.$$

Theorem A.5. Let s and u $\geqq 0$ integers with $0 \leqq s < n$. Let χ be a form of class C^2 and bidegree $(m-s-1, m-s-1)$ on M. Let G be Stokes admissible for χ. Let $\psi: M \longrightarrow \mathbb{R}$ be of class C^2 and have consistent sign around ∂G. Let $f: M \longrightarrow \mathbb{C}^k$ be a holomorphic map of pure rank n. Then

$$\int_G d[(\log |f|)^u f^*(\omega^s) \wedge d^c(\psi\chi)] = \int_{dG} (\log |f|)^u f^*(\omega^s) \wedge d^c\psi \wedge \chi.$$

Proof. Define $\xi(\chi) = (\log |f|)^u f^*(\omega^s) \wedge d^c(\psi\chi)$. Applying Theorem A.4, it suffices to show that $\xi(\chi)$ is integrable over dG. Let $j: dG \longrightarrow \mathbb{R}(M)$ be the inclusion. Then $j^*(d^c(\psi\chi)) = j^*((d^c\psi) \wedge \chi)$. Assume $\psi|G \cap W \leqq 0$ or $\psi|(W-G) \geqq 0$ for some open neighborhood W of G.

If $\chi \geqq 0$, then $j^*(\xi(\chi)) \geqq 0$ by Lemma 7.2. Hence $\xi(\chi)$ is integrable over dG by Theorem A.4. If χ is real, a form $\eta > 0$ of class C^∞ and bidegree $(m-s-1, m-s-1)$ exists such that $\eta + \chi > 0$ on a neighborhood of the compact set $K = \overline{G} \cap \operatorname{supp} \chi$. Then $\eta + \chi > 0$ in a neighborhood of G. Hence $\xi(\eta+\chi)$ and $\xi(\eta)$ are integrable over dG, which shows that $\xi(\chi) = \xi(\eta+\chi) - \xi(\eta)$ is integrable over dG. If χ is complex valued, split χ in real and imaginary parts. If $\psi|G \cap W \geqq 0$ or $\psi|(W-G) \leqq 0$ replace ψ by $-\psi$;

q.e.d.

Theorem A6. (Residue Theorem). Let $f: M \longrightarrow \mathbb{C}^n$ be a q-fibering holomorphic map with $m - n = q \geqq 0$. Define $F = f^{-1}(0)$. Let χ be a form of class C^1 and bidegree (q,q) on M. Let G be Stokes admissible of χ. Assume $f^*(\sigma) \wedge \chi$ is integrable over dG. (See (1.2)). Define $T = \mathrm{supp}_{\partial G}\chi$. Assume $F \cap T$ has measure zero on F. Then

$$\int_{dG} f^*(\sigma) \wedge \chi = \int_{F \cap G} \nu_f \chi - \int_G f^*(\sigma) \wedge d\chi. \tag{2}$$

Proof. Observe $d\sigma = 0$. Define $h_\varepsilon = g_\varepsilon(|f|)$ for a test family $\{g_\varepsilon\}$. Stokes Theorem implies

$$\int_{dG} h_\varepsilon f^*(\sigma) \wedge \chi = \int_G dh_\varepsilon \wedge f^*(\sigma) \wedge \chi - \int_G h_\varepsilon f^*(\sigma) \wedge d\chi.$$

By Lemma A.3, $f^*(\sigma) \wedge d\chi$ is integrable over G. If $\varepsilon \longrightarrow 0$ the Lebesgue's bounded convergence Theorem and Theorem A.1 imply (2), q.e.d.

If φ, ψ, χ have bidegree $(r,r),(p,p),(q,q)$ respectively with $p + q + r = m - 1$, then

$$d\varphi \wedge d^c\psi \wedge \chi = d\psi \wedge d^c\varphi \wedge \chi. \tag{3}$$

Theorem A7. Let $f: M \longrightarrow \mathbb{C}^k$ be a holomorphic map of pure rank n. Take $s \in \mathbb{Z}[0,n-1]$. Define $q = m - s - 1$. Let χ be a form of bidegree (q,q) and class C^2 on M. Let G be Stokes admissible for χ. Assume that $d^c \log |f|^2 \wedge f^*(\omega^s) \wedge \chi$ is integrable over dG. Define $F = f^{-1}(0)$ and $T = \mathrm{supp}_{\partial G}\chi$. If $s = n - 1$, assume that $T \cap F$ has measure zero on F. Assume that $\log |f|^2 f^*(\omega^s) \wedge d^c\chi$ is either integrable over dG or that χ is real and that $f^*(\omega^s) \wedge d^c\chi$ has separate signs on dG. If $s < n - 1$, define

$$L_s(\chi) = \int_G f^*(\omega^{s+1}) \wedge \chi. \tag{4}$$

If $s = n - 1$, assume $k = n$. Then f is q-fibering and

$$L_s(\chi) = \int_{G \cap F} \nu_f \chi \tag{5}$$

exists.

Then

$$\int_{dG} d^c \log |f|^2 \wedge f^*(\omega^s) \wedge \chi + \int_G \log |f|^2 f^*(\omega^s) \wedge dd^c\chi$$

$$= L_s(\chi) + \int_{dG} \log |f|^2 f^*(\omega^s) \wedge d^c\chi$$

and

$$\int_G f^*(\omega^s) \wedge dd^c\chi = \int_{dG} f^*(\omega^s) \wedge d^c\chi. \tag{6}$$

Proof. (6) follows from Theorem A.4 with $u = 0$. Since $K = \overline{G} \cap \operatorname{supp} \chi$ is compact $|f| \leqq 1/e$ on K can be assumed w.ℓ.o.g. Theorem A4, identify (3) and if $s = n - 1$ Theorem 9.6 imply

$$\int_{dG} d^c \log |f|^2 \wedge f^*(\omega^s) \wedge \chi + \int_G \log |f|^2 f^*(\omega^s) \wedge dd^c\chi$$

$$= L_s(\chi) - \int_G d^c \log |f|^2 \wedge f^*(\omega^s) \wedge d\chi$$

$$+ \int_G \log |f|^2 f^*(\omega^s) \wedge dd^c\chi$$

$$= L_s(\chi) + \int_G d(\log |f|^2 f^*(\omega^s) \wedge d^c\chi)$$

$$= L_s(\chi) + \int_{dG} \log |f|^2 f^*(\omega^s) \wedge d^c\chi \qquad \text{q.e.d.}$$

The same method as in the proof of Theorem A5 shows:

Theorem A8. Let $f\colon M \longrightarrow \mathbb{C}^k$ be a holomorphic map of pure rank n. Take $s \in \mathbb{Z}[0,n-1]$. Define $q = m - s - 1$. Let χ be a form of Class C^2 and bidegree (q,q) on M. Let G be Stokes admissible for χ. Let $\psi\colon M \longrightarrow \mathbb{R}$ be of class C^2 with a consistent sign around ∂G. If $s < n - 1$, define $L_s(\chi)$ by (4). If $s = n - 1$, assume $k = n$. Then f is q-fibering and $L_s(\chi)$ is defined by (5). Then

$$\int_G \log |f|^2 f^*(\omega^s) \wedge dd^c(\psi\chi) =$$

$$= L_s(\chi) + \int_{dG} \log |f|^2 f^*(\omega^s) \wedge d^c\psi \wedge \chi.$$

If $s = n - 1$, Theorems A.7 and A.8 are called Green's Residue Theorems. If $s < n - 1$, they are referred to as the Green-Stokes Theorems. For more details to this section see [82] g a) b) c).

b) The double logarithm theorems

If the singularities are weak enough, Stokes Theorem holds without residue. In this section, M is a complex space of pure dimension m. For $x > 0$ define

$$\Omega(x) = \log(\log x)^2 \tag{7}$$

Lemma A9. Let h be a holomorphic function on M such that $h^{-1}(0)$ is thin. Let $w: M \longrightarrow \mathbb{R}$ be a function of class C^∞. Define $v = |h|^2 e^w$. Assume $ev \leqq 1$. Let ξ be a form of class C^1 and degree $2m - 1$ on M. Let G be Stokes admissible for ξ. Assume that either $\Omega(v)\,\xi$ is integrable over dG or that ξ has separate signs on dG. Then

$$\int_{dG} \Omega(v)\,\xi = \int_G d\,\Omega(v) \wedge \xi + \int_G \Omega(v)\,d\,\xi. \tag{8}$$

Proof. By the use of a partition of unity, it suffices to construct an open neighborhood U to each $x_0 \in M$ such that (8) holds if ξ has compact support in U. If $v(x_0) \neq 0$, this is trivial. Hence assume $v(x_0) = 0$. If ξ separates signs, let $j: dG \longrightarrow \mathfrak{R}(M)$ be the inclusion and define

$$\Gamma_+ = \{x \in dG \mid j^*(\xi) > 0\} \qquad \Gamma_- = \{x \in dG \mid j^*(\xi) < 0\}$$

Then $\overline{\Gamma}_+ \cap \overline{\Gamma}_- = \emptyset$. Take an open neighborhood U of x_0 satisfying the following conditions:

(a) The closure $\overline{U}$ is compact with $e|h|^2 < 1$ and $|h|\,e^{|w|} < 1$ on $\overline{U}$. Also h is (m-1)-fibering in a neighborhood of $\overline{U}$.

(b) If ξ has separate signs and if $x_0 \notin \overline{\Gamma}_+ \cup \overline{\Gamma}_-$, then $U \cap (\overline{\Gamma}_+ \cup \overline{\Gamma}_-) = \emptyset$.

(c) If ξ has separate signs and if $x_0 \in \overline{\Gamma}_+$, then $U \cap \overline{\Gamma}_- = \emptyset$.

(d) If ξ has separate signs and if $x_0 \in \overline{\Gamma}_-$, then $U \cap \overline{\Gamma}_+ = \emptyset$.

Take a test family $\{g_\varepsilon\}_{\varepsilon \in \mathbb{R}(0,1)}$. Then $\gamma_\varepsilon = g_\varepsilon(|h|)$ is a function of class C^∞ on M with $0 \leqq \gamma_\varepsilon \leqq 1$. Also $\gamma_\varepsilon = 0$ in a neighborhood of $h^{-1}(0)$. Take ξ with compact support in U. Then Stokes Theorem applies:

$$(9) \qquad \int_{dG} \gamma_\varepsilon \Omega(v)\ \xi = \int_G d\gamma_\varepsilon \wedge \Omega(v)\ \xi + \int_G \gamma_\varepsilon\ d\Omega(v) \wedge \xi + \int_G \gamma_\varepsilon \Omega(v)\ d\xi .$$

Now (a) and $ev \leqq 1$ imply

$$(10) \qquad 1 \leqq \frac{|\log v|}{|\log |h||} \leqq 3$$

on $U - h^{-1}(0)$. Hence $0 \leqq \Omega(v) \leqq 3(-\log|h|^2)$ on U. The forms

$$(11) \qquad \chi = \frac{2\Omega(v)}{-\log|h|^2}\ d\xi \quad \text{and} \quad \chi_0 = \frac{2\Omega(v)}{-\log|h|^2}\ \xi$$

are locally bounded on U and of class C^∞ on $U - h^{-1}(0)$. Define $K = \overline{G} \cap \operatorname{supp} \xi$. Lemma A3 shows the existence of the integral

$$2 \int_K \log \frac{1}{|h|}\ \chi = \int_G \Omega(v)\ d\xi .$$

Observe $\gamma_\varepsilon \longrightarrow 1$ for $\varepsilon \longrightarrow 0$ on $M - h^{-1}(0)$ with $0 \leqq \gamma_\varepsilon \leqq 1$. Therefore

$$(12) \qquad \int_G \gamma_\varepsilon \Omega(v)\ d\xi \longrightarrow \int_G \Omega(v)\ d\xi \qquad \text{for } \varepsilon \longrightarrow 0.$$

Regarding χ_0, Lemma A2 implies

$$(13) \qquad \int_G d\gamma_\varepsilon \wedge \Omega(v)\ \xi = \int_G d\gamma_\varepsilon \wedge \log \frac{1}{|h|}\ \chi_0 \longrightarrow 0$$

for $\varepsilon \longrightarrow 0$. Define

$$\chi_1 = \frac{|h|}{h} \frac{2}{\log v}\ \xi \qquad\qquad \chi_2 = \frac{|h|}{h} \frac{2}{\log v}\ \xi$$

$$\chi_3 = \frac{2}{\log v}\ dw \wedge \xi .$$

Then χ_ρ are locally bounded on M and of class C^∞ on $M - h^{-1}(0)$ with

$$(14) \qquad d\Omega(v) \wedge \xi = \frac{dh}{|h|} \wedge \chi_1 + \frac{d\bar{h}}{|h|} \wedge \chi_2 + \chi_3.$$

By Lemma A.3 $|h|^{-1}\, dh \wedge \chi_1$ and $|h|^{-1}\, d\bar{h} \wedge \chi_2$ are integrable over K, hence over G. Therefore $d\Omega(v) \wedge \xi$ is integrable over G which implies

$$(15) \qquad \int_G \gamma_\varepsilon\, d\Omega(v) \wedge \xi \longrightarrow \int_G d\Omega(v) \wedge \xi \qquad \text{for } \varepsilon \longrightarrow 0.$$

If $\Omega(v)\ \xi$ is integrable over dG, the bounded convergence theorem implies

$$(16) \qquad \int_{dG} \gamma_\varepsilon\, \Omega(v)\ \xi \longrightarrow \int_G \Omega(v)\ \xi \qquad \text{for} \qquad \varepsilon \longrightarrow 0.$$

Consider the case where ξ has separate signs. If (b) holds of U then (16) holds since the integrals are zero. Assume (c). Then $\Omega(v) \leqq 0$ and $j^*(\xi) \geqq 0$ on dG hence $\Omega(v) j^*(\xi) \geqq 0$ on dG. Fatou's Lemma implies

$$0 \leqq \int_{dG} \Omega(v)\ \xi \leqq \lim_{\varepsilon \to 0} \int_{dG} \gamma_\varepsilon \Omega(v)\ \xi = \Lambda$$

where the limit $\Lambda < \infty$ exists by (9), (12), (13) and (15). Hence $\Omega(v)\ \xi$ is integrable over dG and (16) holds. If (d) holds, consider $-\xi$ instead of ξ and (16) follows again. Now, (9), (12), (13), (15) and (16) imply (8), q.e.d.

<u>Lemma A10.</u> Let h be a holomorphic function on M such that $h^{-1}(0)$ is thin. Let $w: M \longrightarrow \mathbb{R}$ be of class C^∞. Define $v = |h|^2 e^w$. Assume $ev \leqq 1$. Let ξ be a real form of degree $2m - 2$ and of class C^1. Let G be Stokes admissible for ξ. Define $T = \operatorname{supp}_{\partial G} \xi$. Assume $T \cap h^{-1}(0)$ has measure zero on $h^{-1}(0)$. Assume that $d^c\Omega(v) \wedge \xi$ is integrable over dG. Then

$$(17) \qquad \int_{dG} d^c\Omega(v) \wedge \xi = \int_G dd^c\Omega(v) \wedge \xi - \int_G d^c\Omega(v) \wedge d\xi.$$

Proof. Again it suffices to construct an open neighborhood U to each $x_0 \in M$ such that (17) holds if ξ has compact support in U. If $v(x_0) \neq 0$, this is trivial. Assume $v(x_0) = 0$. Take an open neighborhood U of x_0 such that $\overline{U}$ is compact, such that $e|h|^2 < 1$ and $|h|e^{|w|} < 1$ on $\overline{U}$ and such that h is (m-1)-fibering in a neighborhood V of $\overline{U}$. Take a test family $\{g_\varepsilon\}_{\varepsilon\in\mathbb{R}(0,1)}$. Then $\gamma_\varepsilon = g_\varepsilon(|h|)$ is a function of class C^∞ on M with $0 \leqq \gamma_\varepsilon \leqq 1$. Also $\gamma_\varepsilon = 0$ in a neighborhood of $h^{-1}(0)$. Take ξ with compact support in U. Stokes Theorem applies:

$$(18)\qquad \int_{dG} \gamma_\varepsilon d^c \varrho(v) \wedge \xi = \int_G d\gamma_\varepsilon \wedge d^c\varrho(v) \wedge \xi + \int_G \gamma_\varepsilon dd^c \varrho(v) \wedge \xi$$

$$- \int_G \gamma_\varepsilon d^c\varrho(v) \wedge d\xi.$$

Here $0 \leqq \gamma_\varepsilon \leqq 1$ and $\gamma_\varepsilon \longrightarrow 1$ for $\varepsilon \longrightarrow 0$ on $M - h^{-1}(0)$. Hence

$$(19)\qquad \int_{dG} \gamma_\varepsilon d^c\varrho(v) \wedge \xi \longrightarrow \int_{dG} d^c\varrho(v) \wedge \xi \qquad \text{for } \varepsilon \longrightarrow 0.$$

Again $\varrho(v) \geqq 0$ and $-\log v \geqq 1$ on M. Also (10) holds on $\overline{U}$. Define $\sigma = d^c \log |z|^2$ on $\mathbb{C} - \{0\}$. Then

$$d^c \log v = d^c \log |h|^2 + d^c w.$$

$$(20)\qquad d^c\varrho(v) = 2\Big(\frac{h^*(\sigma)}{\log v} + \frac{d^c w}{\log v}\Big).$$

Define $\chi_0 = 2\,\xi/\log v$ on $M - h^{-1}(0)$ and $\chi_0 = 0$ on $h^{-1}(0)$. Then χ_0 is a continuous form of bidegree 2m - 2 on M. Define $G_0 = G \cap U$ and $H = G_0 \cap h^{-1}(0)$. Let $j: H \longrightarrow M$ be the inclusion, then $j^*(\chi_0) = 0$. Define $T_0 = \mathrm{supp}_{\partial G}\chi_0 = \mathrm{supp}_{\partial G}\chi_0$. Theorem A.1 applies with the following translation table

Th.A.1	M	m	n	q	f	G	χ	T	F	g_ε	h_ε	τ	σ
Here	V	m	1	m-1	h	G_0	χ_0	T_0	$V \cap h^{-1}(0)$	g_ε	γ_ε	$\lvert z\rvert^2$	σ

Hence

$$(21)\qquad \int_G d\gamma_\varepsilon \wedge \frac{2h^*(\sigma)}{\log v} \wedge \xi = \int_{G_0} d\gamma_\varepsilon \wedge h^*(\sigma) \wedge \chi_0 \longrightarrow \int_H \nu_h \chi_0 = 0$$

for $\varepsilon \longrightarrow 0$. Lemma A2 implies

$$(22)\qquad \int_G d\gamma_\varepsilon \wedge \frac{2d^c w}{\log v} \wedge \xi \longrightarrow 0 \qquad \text{for } \varepsilon \longrightarrow 0.$$

Hence (20),(21) and (22) yield

$$(23)\qquad \int_G d\gamma_\varepsilon \wedge d^c \mathfrak{L}(v) \wedge \xi \longrightarrow 0 \qquad \text{for } \varepsilon \longrightarrow 0.$$

The form $\zeta = |z|\sigma$ has bounded coefficients. Lemma A2 gives the existence of the integral

$$\int_G \frac{2h^*(\sigma)}{\log v} \wedge d\xi = \int_{G_0} \frac{h^*(\zeta)}{|h|} \wedge \frac{2d\xi}{\log v}.$$

Since $(2d^c w \wedge d\xi)/\log v$ is integrable over G, the integral $d^c \mathfrak{L}(v) \wedge d\xi$ exists by (20). Therefore

$$(24)\qquad \int_G \gamma_\varepsilon\, d^c \mathfrak{L}(v) \wedge d\xi \longrightarrow \int_G d^c \mathfrak{L}(v) \wedge d\xi$$

for $\varepsilon \longrightarrow 0$.

Observe $d\sigma = 0$. Hence (20) implies

$$dd^c \mathfrak{L}(v) = \frac{2h^*(\sigma)}{(\log v)^2} \wedge \frac{dv}{v} + \frac{2d^c w}{(\log v)^2} \wedge \frac{dv}{v} + \frac{2dd^c w}{\log v}$$

$$= \frac{2h^*(\sigma)}{(\log v)^2} \wedge d(\log|h|^2) + \frac{2h^*(\sigma)}{(\log v)^2} \wedge dw$$

$$(25)\qquad + \frac{2d^c w}{(\log v)^2} \wedge d(\log |h|^2) + \frac{2d^c w \wedge dw}{(\log v)^2}$$

$$+ \frac{2dd^c w}{\log v}$$

Trivially, the integral

$$\int_G \left(\frac{2d^c w \wedge dw}{(\log v)^2} + \frac{2dd^c w}{\log v}\right) \wedge \xi \tag{26}$$

exists. Lemma A3 implies the existence of the integral

$$\int_G \frac{2h^*(\sigma)}{(\log v)^2} \wedge dw \wedge \xi. \tag{27}$$

Observe

$$2d^c w \wedge d \log |h|^2 \wedge \xi = 2d^c \log |h|^2 \wedge dw \wedge \xi = 2h^*(\sigma) \wedge dw \wedge \xi.$$

Hence, the integral

$$\int_G \frac{2d^c w}{(\log v)^2} \wedge d \log |h|^2 \wedge \xi \tag{28}$$

exists. Let ξ_0 be the projection of ξ into the forms of bidegree $(m-1,m-1)$. Since ξ_0 is real, a positive form $\eta > 0$ of bidegree $(m-1,m-1)$ and class C^∞ exists on U such that $-\eta \leqq \xi_0 \leqq \eta$ on U. Also a function $\lambda: M \longrightarrow \mathbb{R}[0,1]$ of class C^∞ and with compact support in U exists such that $\lambda = 1$ on supp ξ_0. Then $-\lambda\eta \leqq \xi_0 \leqq \lambda\eta$ on U. Also (10) implies

$$\tfrac{1}{4} \leqq \left(\frac{\log |h|^2}{(\log v)}\right) \leqq 4.$$

Observe $h^*(\sigma) \wedge d \log |h|^2 \wedge \xi = - d \log |h|^2 \wedge d^c \log |h| \wedge \xi_0$. Hence the integral

$$\int_G \frac{2dh^*(\sigma)}{(\log v)^2} \wedge d \log |h|^2 \wedge \xi \tag{29}$$

exists if the integral

$$\mathfrak{J} = \int_U \frac{d \log |h|^2 \wedge d^c \log |h|^2}{(\log |H|^2)^2} \wedge \lambda\eta \geqq 0 \tag{30}$$

exists, where $\mathfrak{J}$ has a non-negative integrand. Observe

$$d \log |h|^2 \wedge d^c \log |h|^2 = \frac{i}{2\pi} \quad \frac{dh \wedge d\bar{h}}{|h|^2}$$

Define $A(w) = h^{-1}(w) \cap U$. According to Tung [87] Theorem 5.1.1 the fiber integral

$$F(w) = \int_{A(w)} v_h \, \lambda\eta \geqq 0$$

is a continuous function on $\mathbb{C}$. Define $\theta = 1/\sqrt{e}$. Then $e|h|^2 \leqq 1$ implies $F(w) = 0$ if $|w| > \theta$. A constant $c > 0$ exists such that $F(w) \leqq c$ for all $w \in \mathbb{C}[\theta]$. Therefore

$$0 \leqq \mathfrak{I} = \int_{\mathbb{C}(\theta)} F(w) \frac{i}{2\pi} \frac{dw \wedge d\overline{w}}{|w|^2(\log(w|^2)^2}$$

$$\leqq c \int_{\mathbb{C}(\theta)} \frac{i}{2\pi} \frac{dw \wedge d\overline{w}}{|w|^2(\log|w|^2)^2} = c < \infty.$$

Therefore the integral (29) exists. Now (25), (26), (27), (28), (29) imply that $dd^c\Omega(v) \wedge \xi$ is integrable over G. Hence

(31) $$\int_G \gamma_\varepsilon \, dd^c\Omega(v) \wedge \xi \longrightarrow \int_G dd^c\Omega(v) \wedge \xi \qquad \text{for } \varepsilon \longrightarrow 0.$$

Therefore (18), (19), (23), (24) and (31) imply (17); q.e.d.

For the rest of this section the following assumptions are made

(H) Let M and N be complex spaces. Assume that M has pure dimension m. Let $F\colon M \longrightarrow N$ be a holomorphic map. Let L be a holomorphic line bundle over N with a hermitian metric κ along the fibers of L. Let s be a holomorphic section of L over N such that $f^{-1}(Z(s))$ is thin. Assume $e|s|_\kappa^2 < 1$ on N. Define $v = |s \circ f|_\kappa^2$.

<u>Proposition A11.</u> Assume (H). Let ξ be a form of class C^1 and of degree $2m - 1$ on M. Let G be Stokes admissible for ξ. Assume that either $\Omega(v)\,\xi$ is integrable over dG or that ξ is real and has separate signs on dG. Then

(32) $$\int_{dG} \Omega(v)\,\xi = \int_G d\Omega(v)\,\xi + \int_G \Omega(v)\,d\xi.$$

Proof. Again it suffices to construct an open neighborhood U to each $x_0 \in M$ such that (32) holds if ξ has compact support in U. Take $x_0 \in M$. Take a holomorphic frame u of L over an open neighborhood W of $f(x_0)$. A holomorphic function g on W exists such that $s = g\, u$ on W. Take an open neighborhood U of x_0 such that $f(U) \subseteq W$. Then $h = g \circ f$ is holomorphic on U and $h^{-1}(0) = f^{-1}(Z(s)) \cap U$ is thin in U. The function $w = \log |u \circ f|_\kappa^2$ has class C^∞ on U with $v = |h|^2 e^w$ on U. Suppose that ξ has compact support in U. Then (8) holds for $G_0 = U \cap G$ and (32) follows for G;q.e.d.

By the same method Lemma A10 gives the following result:

Proposition A12. Assume (H). Let ξ be a real form of degree 2m - 2 and of class C^1 on M. Let G be Stokes admissible for ξ. Define $T = \operatorname{supp}_{\partial G} \xi$. Assume that $T \cap f^{-1}(Z(s))$ has measure zero on $f^{-1}(Z(s))$. Assume that $d^c \mathfrak{L}(v) \wedge \xi$ is integrable over dG. Then

$$(33) \qquad \int_{dG} d^c \mathfrak{L}(v) \wedge \xi = \int_G dd^c \mathfrak{L}(v) \wedge \xi - \int_G d^c \mathfrak{L}(v) \wedge d\xi .$$

Theorem A13. Assume (H). Assume a parabolic exhaustion τ on M is given. Take $r \in \mathfrak{C}_\tau$ and $u \in \mathfrak{C}_\tau$ with $0 < u < r$. Then

$$(34) \qquad \int_{M\langle r\rangle} \mathfrak{L}(v)\, \sigma - \int_{M\langle u\rangle} \mathfrak{L}(v)\, \sigma = 2 \int_u^r \int_{M[t]} dd^c \mathfrak{L}(v) \wedge \upsilon^{m-1} \frac{dt}{t^{2m-1}} .$$

Proof. Observe that dM(r) carries the correct orientation as a boundary manifold of the Stokes domain M(r) - M[u] and that dM(u) carries the opposite one. Also σ has separate signs on d(M(r)-M[u]). Proposition A11 implies:

$$\int_{M\langle r\rangle} \mathfrak{L}(v)\, \sigma - \int_{M\langle u\rangle} \mathfrak{L}(v)\, \sigma = \int_{M(u,r]} d\, \mathfrak{L}(v) \wedge d^c \log \tau \wedge w^{m-1}$$

$$= \int_{M(u,r]} d \log \tau \wedge d^c \mathfrak{L}(v) \wedge w^{m-1}$$

$$= 2 \int_u^r \int_{M\langle t\rangle} d^c \mathfrak{L}(v) \wedge \omega^{m-1} \frac{dt}{t}$$

$$= 2 \int_u^r \int_{M\langle t\rangle} d^c \mathfrak{L}(v) \wedge \upsilon^{m-1} \frac{dt}{t^{2m-1}}.$$

The inner integral exists for almost all $t \in \mathbb{R}[u,r]$ and $M\langle t\rangle \cap f^{-1}(Z(s)$ has measure zero on $f^{-1}(Z(s))$ for almost all $t \in \mathbb{R}[u,r]$. Hence Proposition A12 implies

$$(35) \qquad \int_{M\langle t\rangle} d^c \mathfrak{L}(v) \wedge \upsilon^{m-1} = \int_{M[t]} dd^c \mathfrak{L}(v) \wedge \upsilon^{m-1}$$

for almost all $t \in \mathbb{R}[u,r]$, which completes the proof.

Theorem A14. Assume (H). Assume a parabolic exhaustion on τ is given on M. Take $0 < u < r \in \mathbb{R}$. Then

$$(36) \qquad \int_{M(u,r]} dd^c \mathfrak{L}(v) \wedge \omega^{m-1} = r^{2-2m} \int_{M[r]} dd^c \mathfrak{L}(v) \wedge \upsilon^{m-1} - u^{2-2m} \int_{M[u]} dd^c \mathfrak{L}(v) \wedge \upsilon^{m-1}.$$

Proof. Let E be the set of all $t \in \mathfrak{C}_\tau$ such that $M\langle t\rangle \cap f^{-1}(Z(s))$ has measure zero on $f^{-1}(Z(s))$ and such that

$$\int_{M\langle t\rangle} d^c \mathfrak{L}(v) \wedge \upsilon^{m-1} = r^{2m-2} \int_{M\langle t\rangle} d^c \mathfrak{L}(v) \wedge \omega^{m-1}$$

exists. Then $\mathbb{R}_+ - E$ has measure zero as seen in the last proof. If $0 < u < r \in \mathfrak{C}_\tau$ and $u \in \mathfrak{C}_\tau$, Proposition A12 and (35) imply

$$\int_{M(u,r]} dd^c \mathfrak{L}(v) \wedge \omega^{m-1}$$

$$= \int_{M\langle r\rangle} d^c \mathfrak{L}(v)\, \omega^{m-1} - \int_{M\langle u\rangle} d^c \mathfrak{L}(v) \wedge \omega^{m-1}$$

$$= r^{2-2m} \int_{M[r]} dd^c \mathfrak{L}(v) \wedge \upsilon^{m-1} - u^{2-2m} \int_{M[u]} dd^c \mathfrak{L}(v) \wedge \upsilon^{m-1}$$

Semi-continuity from the right proofs the theorem for all $0 < u < r$;

q.e.d.

For the results of this section b) compare Griffiths-King [33] proofs of Lemma to 6.18 and 7.30.

The results of these two sections may be technical but they are fundamental to the derivation of the First and Second Main Theorems. Therefore proofs had to be provided. They are difficult and are based on the deep results of Tung [87].

c) General position

The concept of general position was introduced in section 16a). Here are some additional observations:

Let V be a complex vector space. Then $a_1,\ldots,a_q$ in $\mathbb{P}(V)$ are said to be linearly independent if $a_\mu = \mathbb{P}(\mathfrak{a}_\mu)$ and if $\mathfrak{a}_1,\ldots,\mathfrak{a}_p$ are linearly independent. If so, define

$$(37)\qquad a = a_1 \wedge \ldots \wedge a_p = \mathbb{P}(\mathfrak{a}_1 \wedge \ldots \wedge \mathfrak{a}_p).$$

If $a_1,\ldots,a_q$ in $\mathbb{P}(V)$ and $\mu \in \mathfrak{T}(p,q)$ are given, then a_μ is said to exist if $a_{\mu(1)},\ldots,a_{\mu(p)}$ are linearly independent and if $a_\mu = a_{\mu(1)} \wedge \ldots \wedge a_{\mu(p)}$.

Let L be a holomorphic line bundle over the pure n-dimensional complex manifold N. Let $\eta\colon N \times V \longrightarrow L$ be a simplification. If $a \in \mathbb{P}(V)$, then $a = \mathbb{P}(\mathfrak{a})$ and $s = \eta_{\mathfrak{a}}$ is a section of L with $Z(s) = E_L[a]$. Assume L is adapted to a. Then the divisor θ_s exists. Define $\theta_a = \theta_a[L] = \theta_j$. Observe if ι is the identity, then $\theta_a = \theta_\iota^a[L]$ and θ_a does not depend on the choice of $\mathfrak{a}$. If $a \in G_p(V)$, also put $\theta_a = \theta_\iota^a[L]$.

Proposition A15. Let N, L, V, η as above. Take $a_1,\ldots,a_q$ in $\mathbb{P}(V)$. Then $a_1,\ldots,a_q$ are in general position for L if and only if the following condition is satisfied

($\mathfrak{S}$) Take $p \in \mathbb{N}[1,q]$. Take $\mu \in \mathfrak{T}(p,q)$ with

$$(38)\qquad A = \bigcap_{\lambda=1}^{p} E_L[a_{\mu(\lambda)}] \neq \emptyset.$$

Then $a_\mu \in G_{p-1}(V)$ exists and L is adapted to a_μ with

$$\theta_{a_\mu} | E_L[a_\mu] = 1. \text{ (Observe } E_L[a_\mu] = A).$$

<u>Proof.</u> 1. <u>Assume that $a_1,\ldots,a_q$ are in general position.</u>

Take $p \in \mathbb{N}[1,q]$. Take $\mu \in \mathfrak{T}(p,q)$. Define A by (38) and assume $A \neq \emptyset$. Take $x \in A$. Then $r \in \mathbb{N}[1,q]$ with $p \leqq r \leqq q$ and $\rho \in \mathfrak{T}(r,q)$ exist such that $\rho|\mathbb{N}[1,p] = \mu$ and such that $x \in E_L[a_{\rho(1)}] \cap \ldots \cap E_L[a_{\rho(r)}]$ and such that $x \notin E_L[a_j]$ if $j \in \mathbb{N}[1,q] - \operatorname{Im}\rho$. There exist holomorphic functions $f_1,\ldots,f_r$ on an open, connected neighborhood U of x, such that $f_\lambda \not\equiv 0$ on U and such that $\theta_{f_\lambda} = \theta_{a_{\rho(\lambda)}}|U$ for $\lambda = 1,\ldots,r$ and such that $(df_1\wedge\ldots\wedge df_r)(z) \neq 0$ for all $z \in U$. Also U can be taken such that a holomorphic frame v of L over U exists. Define $\mathfrak{w}: U \longrightarrow V^*$ by (3.4). Take $0 \neq \mathfrak{a}_j \in \mathbb{P}(V)$ with $\mathbb{P}(\mathfrak{a}_j) = a_j$. Define $w_\lambda = (\mathfrak{w}, \mathfrak{a}_{\rho(\lambda)})$ for $\lambda = 1,\ldots,r$. Then $w_\lambda^{-1}(0) = E_L[a_{\rho(\lambda)}] \cap U$ and $\theta_{w_\lambda} = \theta_{a_{\rho(\lambda)}}|U = \theta_{f_\lambda}$.

Hence holomorphic functions $g_\lambda: U \longrightarrow \mathbb{C} - \{0\}$ exist on U such that $w_\lambda = g_\lambda f_\lambda$. Then

$$(dw_1\wedge\ldots\wedge dw_r)(x) = g_1(x)\ldots g_r(x)(df_1\wedge\ldots\wedge df_r)(x) \neq 0.$$

Hence U can be taken so small that $(dw_1\wedge\ldots\wedge dw_r)(z) \neq 0$ for all $z \in U$. Assume $a_{\mu(1)},\ldots,a_{\mu(p)}$ are linearly dependent. Then $\mathfrak{a}_{\mu(1)},\ldots,\mathfrak{a}_{\mu(p)}$ and consequently $w_1,\ldots w_p$ are linearly dependent. Therefore $dw_1\wedge\ldots\wedge dw_p \equiv 0$ which is impossible. Therefore $a_\mu = a_{\mu(1)}\wedge\ldots\wedge a_{\mu(p)}$ exists. The holomorphic map $w = (w_1,\ldots,w_p)$ is regular. Hence $w^{-1}(0) = E_L[a_\mu] \cap U$ has pure dimension $(n-p)$ and L is adapted to a_μ at x. Also $1 = \nu_w(x) = \theta_{a_\mu}(x)$ by Tung [87] Lemma 2.2.2.2 and by definition of $\theta_{a_\mu} = \theta_\iota^{a_\mu}[L]$. Condition $(\mathfrak{J})$ is satisfied.

2. <u>Assume $(\mathfrak{J})$ is satisfied.</u> Take $x \in N$. Assume $x \in E_L[a_j]$ for at least one $j \in \mathbb{N}[1,q]$. Then there exist uniquely $r \in \mathbb{N}[1,p]$ and $\mu \in \mathfrak{T}(r,q)$ such that $x \in E_L[a_{\mu(\lambda)}]$ for all $\lambda = 1,\ldots,r$ and that

$x \notin E_L[a_j]$ if $j \notin \operatorname{Im} \mu$. Take an open, connected neighborhood U of x such that $U \cap E_L[a_j] = \emptyset$ if $j \notin \operatorname{Im} \mu$ and such that a holomorphic frame $\mathfrak{v}$ of L over U exists. Define $\mathfrak{w}: U \longrightarrow V^*$ by (3.4). Take $0 \neq \mathfrak{a}_j \in V$ with $\mathbb{P}(\mathfrak{a}_j) = a_j$. Define $w_\lambda = (\mathfrak{w}, \mathfrak{a}_{\mu(\lambda)})$ and $w = (w_1, \ldots, w_r)$. By assumption a_μ exists with $\theta_w = \theta_{a_\mu} = 1$ on $U \cap E_L[a_\mu]$ and $\theta_\mu = \theta_{a_\mu} = 0$ on $U - E_L[a_\mu]$. By Tung [87] 2.2.2.2 w is regular at x. If U is small enough $(dw_1 \wedge \ldots \wedge dw_r)(z) \neq 0$ for all $z \in U$. Condition (G) of 16a) is satisfied and $a_1, \ldots, a_q$ are in general position; q.e.d.

The following proposition shows that the concept of general position coincides with the classical concept for the hyper-plane section bundle.

Proposition A16. Let V be a complex vector space of dimension $n + 1$. Let $S_0(V)^*$ be the hyper-plane section bundle. Let $\eta: \mathbb{P}(V) \times V^* \longrightarrow S_0(V)^*$ be the standard amplification. Take $a_1, \ldots, a_q$ in $\mathbb{P}(V^*)$. Then $a_1, \ldots, a_q$ are in general position if and only if the following property holds:

(P) Take any $p \in \mathbb{N}[1,q]$ with $1 \leqq p \leqq n + 1$. Take any $\mu \in \mathfrak{T}(p,q)$. Then $a_{\mu(1)}, \ldots, a_{\mu(p)}$ are linearly independent.

Proof. 1. Assume $a_1, \ldots, a_q$ are in general position. Take $p \in \mathbb{N}[1,q]$ with $1 \leqq p \leqq n + 1$. Take $\mu \in \mathfrak{T}(p,q)$. If $1 \leqq p \leqq n$, then $\ddot{E}[a_{\mu(1)}], \ldots, \ddot{E}[a_{\mu(p)}]$ intersect. Hence $a_{\mu(1)}, \ldots, a_{\mu(p)}$ are linearly independent by Proposition A15. Now assume $p = n + 1$. Then $a_{\mu(1)}, \ldots, a_{\mu(n)}$ are linearly independent and $x \in \ddot{E}[a_{\mu(1)}] \cap \ldots \cap \ddot{E}[a_{\mu(n)}]$ exists. If $a_{\mu(1)}, \ldots, a_{\mu(n+1)}$ are linearly dependent, then $x \in \ddot{E}[a_{\mu(n+1)}]$. Now (16.1) implies $p \leqq n$. Contradition! Hence $a_{\mu(1)}, \ldots, a_{\mu(n+1)}$ are linearly independent.

2. Assume condition (P) is satisfied. Take $x \in \mathbb{P}(V)$. Define $p = \#\{\lambda \in \mathbb{N}[1,q] \mid x \in \ddot{E}[a_\lambda]\}$. Then $p \leqq n$ is claimed. Assume $p > n$. Then $\mu \in \mathfrak{T}(n+1,q)$ exists such that $x \in \ddot{E}[a_{\mu(\lambda)}]$ for all $\lambda = 1, \ldots, n + 1$.

Hence $a_{\mu(1)},\ldots,a_{\mu(n+1)}$ are linearly independent by (P). Hence $\ddot{E}[a_{\mu(1)}]\wedge\ldots\wedge\ddot{E}[a_{\mu(n+1)}] = \emptyset$. Contradition! Therefore $p \leqq n$. If $p = 0$, nothing is to be shown. Assume $p \geqq 1$. Take $\mu \in \mathfrak{T}(p,q)$ such that $x \in \ddot{E}[a_{\mu(\lambda)}]$ for all $\lambda = 1,\ldots,p$ and $x \notin \ddot{E}[a_\nu]$ for all $\nu \notin \mathrm{Im}\,\mu$. A base $\beta_1,\ldots,\beta_{n+1}$ of V^* exists such that $\mathbb{P}(\beta_\lambda) = a_{\mu(\lambda)}$ for $\lambda = 1,\ldots,p$. Define $\beta = \beta_{n+1}$ and $b = \mathbb{P}(\beta)$. Then $x \in \mathbb{P}(V) - \ddot{E}[b]$. An open neighborhood U of x in $\mathbb{P}(V) - \ddot{E}[b]$ exists such that $U \cap \ddot{E}[a_\nu] = \emptyset$ if $\nu \notin \mathrm{Im}\,\mu$. Holomorphic functions $w_1,\ldots,w_n$ exist in $\mathbb{P}(V) - \ddot{E}[b]$ such that $w_\mu \circ \mathbb{P} = \beta_\mu/\beta$ on $V - E[b]$ for $\mu = 1,\ldots,n$. Then

$$w = (w_1,\ldots,w_n) : \mathbb{P}(V) - \ddot{E}[b] \longrightarrow \mathbb{C}^n$$

is a patch on $\mathbb{P}(V)$. Hence $(dw_1\wedge\ldots\wedge dw_p)(z) \neq 0$ for all $z \in U$. Also $w_\lambda^{-1}(0) = \ddot{E}[a_{\mu(\lambda)}] - \ddot{E}[b]$ for $\lambda = 1,\ldots,p$ and $\theta_{w_\lambda} = 1$ if $w_\lambda(z) = 0$ condition (G) of section 16 is satisfied and $a_1,\ldots,a_q$ are in general position;q.e.d.

List of General Assumptions

A) General Assumptions A:

(A1) Let M be a complex space of pure dimension $m > 0$.

(A2) Let N be a complex space.

(A3) Let L be a holomorphic line bundle over N with a hermitian metric κ along the fibers of L.

(A4) Let V be a complex vector space of dimension $k + 1 \geqq 1$ with a hermitian metric ℓ.

(A5) Let $\eta: N \times V \longrightarrow L$ be a semi-amplification of L.

(A6) Let $f: M \longrightarrow N$ be meromorphic map.

(A7) Let (G, g, ψ) be a condensor.

(A8) Take $p \in \mathbb{Z}[0,k]$ with $q = m - p$.

(A9) Let $\chi \geqq 0$ be a non-negative form of bidegree (q,q) and class C^2 on M such that $d\chi = 0$.

B) General Assumptions B:

(B1) Let τ be a logarithmic pseudo-convex exhaustion of the complex space M of pure dimension $m > 0$.

(B2) Let L be a holomorphic line bundle over the complex space N. Let κ be a hermitian metric along the fibers of L.

(B3) Let V be a complex vector space of dim $k + 1 \geqq 1$ with a hermitian metric ℓ.

(B4) Let $\eta: N \times V \longrightarrow L$ be a semi-amplification of L.

(B5) Let $f: M \longrightarrow N$ be a meromorphic map.

(B6) Take $p \in \mathbb{Z}[0,k]$ with $q = m - p \geqq 0$.

(B7) Assume that $c(L, \kappa) \geqq 0$.

(B8) Assume that κ is distinguished.

(B9) Assume that f is safe for L.

(B10) The meromorphic map f is safe of order p for L.

(B11) The meromorphic map f is almost adapted of order p - 1 for L.

(B12) The exhaustion τ is strong.

C) General Assumptions C:

(C1) Let N be a compact, connected, complex manifold of dimension $n > 0$.

(C2) Let $L_1, \ldots, L_q$ be holomorphic line bundles on N and let κ_j be a hermitian metric along the fibers of L_j for $j = 1, \ldots, q$.

(C3) Let V_j be a linear subspace of $\Gamma(N, L_j)$, let ℓ_j be a hermitian metric on V_j and let $\eta_j: N \times V_j \longrightarrow L_j$ be the evaluation map for $j = 1, \ldots, q$. Assume $\dim V_j > 0$ and that κ_j is distinguished.

(C4) Given are $a_j \in \mathbb{P}(V_j)$ for $j = 1, \ldots, q$ such that $a_1, \ldots, a_q$ are in general position. Define $u_j = \|\Box, a_j\|^2_{\kappa_j}$ for $j = 1, \ldots, q$.

(C5) Define $L = L_1 \otimes \ldots \otimes L_q$ and $\kappa = \kappa_1 \otimes \ldots \otimes \kappa_q$ as a hermitian metric along the fibers of L. Define $V = V_1 \otimes \ldots \otimes V_q \subseteq \Gamma(N, L)$. Then $\ell = \ell_1 \otimes \ldots \otimes \ell_q$ is a hermitian metric on V. Define $a = a_1 \otimes \ldots \otimes a_q \in \mathbb{P}(V)$. Let $\eta: V \longrightarrow L$ be the evaluation map.

(C6) Let $\Omega > 0$ be a form of class C^∞ and degree 2n on N. Let $\lambda > 0$ be given and assume $\lambda\, c(L, \kappa) + \operatorname{Ric} \Omega > 0$.

(C7) Let ξ_a be a Carlson-Griffiths form for the λ of (C6).

(C8) Let τ be a parabolic exhaustion of the connected, complex manifold M of dimension m.

(C9) Let $f: M \longrightarrow N$ be a holomorphic map.

(C10) Assume that f is adapted to a_j for L_j for $j = 1, \ldots, q$.

(C11) Let F be an effective Jacobian section of f.

(C12) The effective Jacobian section F of f is dominated by τ with dominator Y.

D) General Assumptions D:

(D1) Let N be a compact, connected, complex manifold of dimension $n > 0$.

(D2) Let L be a positive, holomorphic line bundle over N.

(D3) Let V be a linear subspace of $\Gamma(N,L)$ with $0 < \dim V = k + 1$. Let ℓ be a hermitian metric on V. Let $\eta: N \times V \longrightarrow L$ be the evaluation map.

(D4) Let κ be a hermitian metric along the fibers of L such that $c(L,\kappa) > 0$. Assume κ is distinguished for η and ℓ.

(D5) Given are $a_1,\ldots,a_q$ in $\mathbb{P}(V)$ such that $a_1,\ldots,a_q$ are in general position for L.

(D6) Let M be a connected complex manifold of dimension $m > 0$.

(D7) Let $f: M \longrightarrow N$ be a holomorphic map.

(D8) Let τ be a parabolic exhaustion of M.

(D9) Let F be an effective Jacobian section of f dominated by τ. Let Y be the dominator.

E) General Assumptions E:

(E1) Let W be a hermitian vector space of dimension $k > 0$. Define τ_W by $\tau_W(\mathfrak{z}) = |\mathfrak{z}|^2$.

(E2) Let M be a m-dimensional, irreducible, affine algebraic variety embedded into W with $0 < m < k$.

(E3) Let S be a projective plane of dimension $k - m - 1$ in $\mathbb{P}(W)$ such that $S \cap \overline{M} = \emptyset$ if $\overline{M}$ is the closure of M in $\overline{\overline{W}}$.

(E4) Let Y be a m-dimensional linear subspace of W such that $\overline{\overline{Y}} \cap S = \emptyset$. Let $\gamma_Y^S: (\overline{\overline{W}}-S) \longrightarrow \overline{\overline{Y}}$ be the projection from center S onto $\overline{\overline{Y}}$. Define $\tilde{\gamma} = \gamma_Y^S \mid \overline{M}$ and $\tilde{\gamma} = \gamma \mid M$.

(E5) Define $\tau_Y = \tau_W \mid Y$ and $\tau = \tau_Y \circ \gamma: M \longrightarrow \mathbb{R}_+$.

(E6) Let N be a connected, compact, complex manifold of dimension $n > 0$.

(E7) Let L be a holomorphic line bundle on N with a hermitian metric κ along the fibers of L such that $c(L,\kappa) > 0$.

(E8) Let $f: M \longrightarrow N$ be a holomorphic map.

REFERENCES

[1] Ahlfors, L.: The theory of meromorphic curves. Acta Soc. Sci. Fenn. Nova Ser. A 3 (4) (1941) pp. 31.

[2] Andreotti, A. and W. Stoll: Analytic and algebraic dependence of meromorphic functions. Lecture Notes in Mathematics 234. Springer-Verlag Berlin-Heidelberg-New York, 1971, pp. 390.

[3] Bloom, T. and M. Herrera: De Rham cohomology of an analytic space. Invent. Math 7 (1969), 275-296.

[4] Bott, R. and S. S. Chern: Hermitian vector bundles and the equidistribution of the zeroes of their holomorphic sections. Acta Math 114 (1965), 71-112.

[5] Carlson, J.: Some degeneracy theorems for entire functions with values in an algebraic variety. Transactions Amer. Math. Soc. 168 (1972), 273-301.

[6] Carlson, J.: A remark on the transcendental Bezout Problem. Value-Distribution Theory Part A (Edited by R.O. Kujala and A.L.Vitter III) Pure and Appl. Math 25 Marcel Dekker New York 1974, 133-143.

[7] Carlson, J.: A result on the value distribution of holomorphic maps $f: \mathbb{C}^n \to \mathbb{C}^n$. Proceed.of Symp.pure Math. 30 (1977) 225-227

[8] Carlson, J. and Ph. Griffiths: A defect relation for equidimensional holomorphic mappings between algebraic varieties. Ann. of Math. (2) 95 (1972), 557-584.

[9] Carlson, J. and Ph. Griffiths: The order functions for entire holomorphic mappings. Value-Distribution Theory Part A. (Edited by R. O. Kujala and A. L. Vitter III) Pure and Appl. Math. 25 Marcel Dekker, New York 1974, 225-248.

[10] Chern, S. S.: The integrated form of the first main theorem for complex analytic mappings in several variables Ann. of Math (2) 71 (1960), 536-551.

[11] Chern, S. S.: Holomorphic curves in the plane. Diff. Geom. in honor of K. Yano, Kinokuniya, Tokyo, 1972, 72-94.

[12] Chern, S. S. - M. Cowen - A. Vitter III: Frenet frames along holomorphic curves. Value Distribution Theory Part A (Edited by R. O. Kujala and A. A. Vitter III) Pure and Appl. Math. 25 Marcel Dekker, New York, 1974, 191-203.

[13] Chow, W.-L.: On compact analytic varieties. Amer. J. Math. 71 (1949), 893-914.

[14] Cornalba, M. and Ph. Griffiths: Analytic cycles and vector bundles on non-compact algebraic varieties. Invent. Math. 28 (1975), 1-106.

[15] Cornalba, M. and B. Shiffmam: A counter example to the "Transcendental Bezout Problem". Ann. of Math (2) 96 (1972), 402-406.

[16] Cowen, M.: Hermitian vector bundles and value distribution for Schubert cycles. Trans. Amer. Math. Soc. 180 (1973), 189-228.

[17] Cowen, M.: The Kobayashi metric on $\mathbb{P}_n - (2^n+1)$ - hyperplanes. Value-Distribution Theory Part A (Edited by R.O. Kujala and A. L. Vitter III Pure and Appl. Math. 25 Marcel Dekker, New York, 1974, 205-223.

[18] Cowen, M. and Ph. Griffiths: Holomorphic curves and metrics of negative curvature. J. Analyse Math. 29 (1976) 93-152.

[19] Dektjarev, L.: The general first fundamental theorem of value distribution. Dokl. Akad. Nauk. SSR 193 (1970) (Soviet Math Dokl. 11 (1970) 961-963).

[20] Drouilhet, S.: A unicity theorem for equidimensional maps. Rice University Thesis 1974. 69 pp. of ms.

[21] Federer, H.: Geometric measure theory. Die Grundl. d. Math. Wiss. 153 Springer-Verlag Berlin-Heidelberg - New York 1969 pp. 1-676.

[22] Grauert, H.: Über Modifikationen und exceptionelle analytische Mengen. Math. Ann. 146 (1962) 331-368.

[23] Green, M.: Holomorphic maps into complex projective space omitting hyperplanes. Transactions Amer. Math Soc. 169 (1972) 89-103.

[24] Green, M.: Some Picard theorems for holomorphic maps to algebraic varieties. Am. J. Math. 97(1975) 43-75.

[25] Green, M.: Some examples and counter-examples in value-distribution theory for several variables. Compositio Math. 30 (1975) 317-322.

[26] Griffiths, Ph.: Holomorphic mapping into canonical algebraic varieties. Ann. of Math. (2) 93 (1971), 439-458.

[27] Griffiths, Ph.: Two theorems on extensions of holomorphic mappings. Invent. Math. 14 (1971) 27-62.

[28] Griffiths, Ph.: Function theory of finite order on algebraic varieties I. Journ. of Diff. Geom. 6 (1972) 285-306. II Ibid 7 (1972) 45-66.

[29] Griffiths, Ph.: Some remarks on Nevanlinna theory. Value-Distribution Theory Part A (Edited by R.O.Kujala and A. L. Vitter III). Pure and Appl. Math. 25 Marcel Dekker, New York 1974, 1-11.

[30] Griffiths, Ph.: Two theorems in the global theory of holomorphic mappings. Contributions to Analysis. A collection of papers dedicated to Lipman Bers. Academic Press (1974), 169-183.

[31] Griffiths, Ph. : On the Bezout problem for entire analytic sets. Annals of Math. (2) 100 (1974) 533-552.

[32] Griffiths, Ph.: Entire holomorphic mappings in one and several complex variables. Annals of Math. Stud. No. 85. Princeton Univ. Press 1976, X + 99 pp.

[33] Griffiths, Ph. and J. King: Nevanlinna theory and holomorphic mappings between algebraic varieties. Acta. Math. 130 (1973), 145-220.

[34] Harvey, R. and A. Knapp: Positive (p,p) forms, Wirtinger's inequality and currents. Value-Distribution Theory Part A (Edited by R. O. Kujala and A. L. Vitter III) Pure and Appl. Math 25 Marcel Dekker, New York 1974, 43-62.

[35] Hirschfelder, J.: The first main theorem of value distribution in several variables. Invent. Math 8 (1969) 1-33.

[36] Kiernan, P. and S. Kobayashi: Holomorphic mappings into projective space with lacunary hyperplanes. Nagoya Math. J. 50 (1973), 199-216.

[37] King, J.: The currents defined by analytic varieties. Acta. Math. 127 (1971), 185-220.

[38] King, J.: Redefined Residues, Chern Forms and Intersections. Value-Distribution Theory Part A (Edited by R. O. Kujala and A. L. Vitter III) Pure and Appl. Math. 25 Marcel Dekker, New York 1974, 169-190.

[39] Kneser, H.: Zur Theorie der gebrochenen Funktionen mehrerer Veränderlichen. Jber. dtsch. Math. Ver. 48 (1938), 1-28.

[40] Kobayashi, S. and T. Ochiai: Mappings into compact complex manifolds with negative first Chern Class. Journ. Math. Soc. Jap. 23 (1971), 137-148.

[41] Kobayashi, S. and T. Ochiai: Mappings into compact complex manifolds with non-negative Chern class. J. Math. Soc. Japan 23 (1971) 137-148.

[42] Kodaira, K.: On Kähler varieties of restricted type (an intrinsic characterization of algebraic varieties). Ann. of Math. 60 (1954) 28-48.

[43] Kodaira, K.: Holomorphic mappings of polydiscs into compact complex manifolds. J. Diff. Geom. 6 (1971), 33-46.

[44] Lelong, P.: Intégration sur une ensemble analytique complexe. Bull. Soc. Math. France 85 (1957) 328-370.

[45] Lelong, P.: Founctions plurisousharmoniques et formes differentielles positive. Gordon and Breach (1968) pp. 301.

[46] Levine, H.: A theorem on holomorphic mappings into complex projective space. Ann. of Math. (2) 71 (1960) 529-535.

[47] Matsushima, Y.: On a problem of Stoll concerning a cohomology map from a flag manifold into a Grassmann manifold. Osaka J. Math. 13 (1976), 231-269.

[48] Murray, J.: A second main theorem of value distribution theory on Stein manifolds with pseudoconvex exhaustion. (1974 Notre Dame Thesis) pp. 69.

[49] Nevanlinna, R.: Eindeutige analytische Funktionen. Die Grundl. d. Math. Wiss. XLVC Springer Verlag. Berlin-Göttingen-Heidelberg 2 ed. 1953 pp. 379.

[50] Noguchi, J.: A relation between order and defects of meromorphic mappings of $\mathbb{C}^n$ into $\mathbb{P}^n(\mathbb{C})$ (13 pp. of ms.).

[51] Noguchi, J.: On meromorphic mappings of finite analytic coverings over $\mathbb{C}^m$ into projective algebraic varieties and the second main theorem of Griffiths-King (9 pp. of ms.)

[52] Noguchi, J.: Meromorphic mappings of a covering space over $\mathbb{C}^m$ into a projective algebraic variety and defect relations. Hiroshima Math. J. 6 (1976) No. 2 , 265-280.

[53] Noguchi, J.: Holomorphic mappings into closed Riemann surfaces. Hiroshima Math. J. 6 (1976) No. 2, 281-291.

[54] Ochiai, T.: Some remarks on the defect relation of holomorphic curves. Osaka J. Math. 11 (1974) 483-501.

[55] Remmert, R.: Holomorphe und meromorphe Abbildungen komplexer Räume. Math. Ann. 133 (1957) 338-370.

[56] Rutishauser, H.: Über Folgen und Scharen von analytischen und meromorphen Funktionen mehrerer Variabeln, sowie von analytischen Abbildungen, Acta. Math. 83 (1950), 249-325.

[57] Sakai, F.: Degeneracy of holomorphic maps with Ramification. Invent. Math. 26 (1974) 213-229.

[58] Sario, L - M. Nakai: Classification Theory of Riemann surfaces. Die Grundl. d. Math. Wiss. 164. Springer-Verlag Berlin-Heidelberg-New York. 1970 pp. 446.

[59] Schwartz, M. - H.: Formules apparenté es a la formule de Gauss-Bonnet pour certaines applications d'une variete a n dimensions dans une autre. Acta. Math. 91 (1954) 189-244.

[60] Schwartz, M. - H.: Formules apparenté es a celles de Nevanlinna-Ahlfors pour certaines applications d'une variete a n-dimensions dans une autre. Bull. Soc. Math. France 82 (1954), 317-360.

[61] Shiffman, B.: Extension of positive line bundles and meromorphic maps. Invent. Math. 15 (1972) 332-347.

[62] Shiffman, B.: Applications of geometric measure theory to value distribution theory for meromorphic maps. Value-Distribution Theory Part A (Edited by R. O. Kujala and A. L. Vitter III) Pure and Appl. Math. 25 Marcel Dekker, New York 1974, 63-95.

[63] Shiffman, B.: Nevanlinna defect relations for singular divisors. Inventiones. Math. 31 (1975) 155-182.

[64] Shiffman, B.: Holomorphic curves in algebraic manifolds (to appear Bull. Amer. Math Soc.)

[65] Stein, K.: Maximale holomorphe und meromorphe Abbildungen I. Amer. J. Math. 85 (1963) pp. 298-313, II ibid. 86 (1964) pp. 823-868.

[66] Stoll, W.: Mehrfache Integrale auf komplexen Mannigfaltigkeiten. Math. Zeitschr. 57 (1952) 116-154.

[67] Stoll, W.: Die beiden Hauptsätze der Wertverteilungstheorie bei Funktionen mehrerer komplexer Veränderlichen. I Acta Math. 90 (1953) 1-115 II. Acta Math. 92 (1954). 55-169.

[68] Stoll, W.: Einige Bemerkungen zur Fortsetzbarkeit analytischer Mengen. Math. Zeitshr. 60 (1954) 287-304.

[69] Stoll, W.: Über die Fortsetzbarkeit analytischer Mengen endlichen Oberflächeninhalts. Arch. d Math. 9 (1958), 167-175.

[70] Stoll, W.: Über meromorphe Abbildungen komplexer Räume I. Math. Ann. 136 (1958) p. 201-239 II. ibid. 136 (1958) 393-429.

[71] Stoll, W.: The growth of the area of a transcendental analytic set I. Math. Ann. 156 (1964) 47-78, II. Math. Ann. 156 (1964) 144-170.

[72] Stoll, W.: Normal families of non-negative divisors. Math. Zeitschr. 84 (1964) 154-218.

[73] Stoll, W.: The multiplicity of a holomorphic map. Invent. Math. 2 (1966), 15-58.

[74] Stoll, W.: A general first main theorem of value distribution. Acta Math 118 (1967) 111-191.

[75] Stoll, W.: About value distribution of holomorphic maps into projective space. Acta Math. 123 (1969) 83-114.

[76] Stoll, W.: Value districution of holomorphic maps into compact complex manifolds. Lecture Notes in Mathematics 135 Springer-Verlag. Berlin-Heidelberg-New York 1970 1-267.

[77] Stoll,W.: Value distribution of holomorphic maps. Several Complex Variables I. Maryland 1970 Lecture Notes in Mathematics 155 (1970) p. 165-170. Springer-Verlag. Berlin-Heidelberg-New York.

[78] Stoll, W.: Fiber integration and some of its application Symposium on Several Complex Variables. Park City, Utah, 1970. Lecture Notes in Mathematics 184 (1971) 109-120. Springer-Verlag. Berlin-Heidelberg-New York.

[79] Stoll, W.: A Bezout estimate for complete intersections. Ann. of Math. (2) 96 (1972) p. 361-401.

[80] Stoll, W.: Deficit and Bezout estimates. Value-Distribution Theory Part B (Edited by R. O. Kujula and A. L. Vitter III) Pure and Appl. Math. 25 Marcel Dekker, New York 1973, 1-271.

[81] Stoll, W.: Value distribution on parabolic spaces. Proceed. of Symp. Pure Math. 30 (1976) 259-263.

[82] Stoll, W.: Value distribution on parabolic spaces. Lecture Notes 1975. 650 pp of ms.

[83] Stoll, W.: Aspects of value distribution theory in several complex variables. Bull. Amer. Math.Soc.83 (1977) 166-183.

[84] Stoll, W.: A Casorati-Weierstrass Theorem for Schubert zeroes in semi-ample holomorphic vector bundles. 46 pp. of ms.

[85] Thie, P.: The Lelong number of point of a complex analytic set Math. Ann. 172 (1967) 269-312.

[86] Thullern, P.: Über die wesentlichen Singularitäten analytischer Funktionen and Flächen im Raum von n komplexen Veränderlichen. Math. Ann. 111, (1934), 137-157.

[87] Tung, Ch.: The first main theorem on complex spaces. (1973) Notre Dame Thesis) pp. 320.

[88] Wells, R. O.: Differential analysis on complex manifolds. Prentice Hall, Englewood Cliffs 1973 pp. 252.

[89] Weyl, H. and J. Weyl: Meromorphic functions and analytic curves Ann. of Math. Studies 12. Princeton Univ. Press, Princeton, N.J. 1943 , 1-269.

[90] Wong, P.: Defect relations for meromorphic maps from parabolic manifolds into complex projective spaces. 239 pp. of ms. (1976 Notre Dame Thesis).

[91] Wu, H.: Mappings of Riemann surfaces (Nevanlinna Theory). Proceed. of Symposia in Pure Math. 11 (1968) 480-532.

[92] Wu, H.: Remarks on the first main theorem of equidistribution theory 1 Jour. of Diff. Geom.2(1968) 197-202 II ibid 2 (1968) 369-384, III ibid 3 (1969) 83-94 IV ibid (1969) 433-446.

[93] Wu, H.: The equidistribution theory of homomorphic curves. Annals. of Math. Studies 64 Princeton Univ. Press. Princeton, N.J. 1970 pp. 219.

INDEX

Vol. 460: O. Loos, Jordan Pairs. XVI, 218 pages. 1975.

Vol. 461: Computational Mechanics. Proceedings 1974. Edited by J. T. Oden. VII, 328 pages. 1975.

Vol. 462: P. Gérardin, Construction de Séries Discrètes p-adiques. »Sur les séries discrètes non ramifiées des groupes réductifs déployés p-adiques«. III, 180 pages. 1975.

Vol. 463: H.-H. Kuo, Gaussian Measures in Banach Spaces. VI, 224 pages. 1975.

Vol. 464: C. Rockland, Hypoellipticity and Eigenvalue Asymptotics. III, 171 pages. 1975.

Vol. 465: Séminaire de Probabilités IX. Proceedings 1973/74. Edité par P. A. Meyer. IV, 589 pages. 1975.

Vol. 466: Non-Commutative Harmonic Analysis. Proceedings 1974. Edited by J. Carmona, J. Dixmier and M. Vergne. VI, 231 pages. 1975.

Vol. 467: M. R. Essén, The Cos $\pi\lambda$ Theorem. With a paper by Christer Borell. VII, 112 pages. 1975.

Vol. 468: Dynamical Systems – Warwick 1974. Proceedings 1973/74. Edited by A. Manning. X, 405 pages. 1975.

Vol. 469: E. Binz, Continuous Convergence on C(X). IX, 140 pages. 1975.

Vol. 470: R. Bowen, Equilibrium States and the Ergodic Theory of Anosov Diffeomorphisms. III, 108 pages. 1975.

Vol. 471: R. S. Hamilton, Harmonic Maps of Manifolds with Boundary. III, 168 pages. 1975.

Vol. 472: Probability-Winter School. Proceedings 1975. Edited by Z. Ciesielski, K. Urbanik, and W. A. Woyczyński. VI, 283 pages. 1975.

Vol. 473: D. Burghelea, R. Lashof, and M. Rothenberg, Groups of Automorphisms of Manifolds. (with an appendix by E. Pedersen) VII, 156 pages. 1975.

Vol. 474: Séminaire Pierre Lelong (Analyse) Année 1973/74. Edité par P. Lelong. VI, 182 pages. 1975.

Vol. 475: Répartition Modulo 1. Actes du Colloque de Marseille-Luminy, 4 au 7 Juin 1974. Edité par G. Rauzy. V, 258 pages. 1975. 1975.

Vol. 476: Modular Functions of One Variable IV. Proceedings 1972. Edited by B. J. Birch and W. Kuyk. V, 151 pages. 1975.

Vol. 477: Optimization and Optimal Control. Proceedings 1974. Edited by R. Bulirsch, W. Oettli, and J. Stoer. VII, 294 pages. 1975.

Vol. 478: G. Schober, Univalent Functions – Selected Topics. V, 200 pages. 1975.

Vol. 479: S. D. Fisher and J. W. Jerome, Minimum Norm Extremals in Function Spaces. With Applications to Classical and Modern Analysis. VIII, 209 pages. 1975.

Vol. 480: X. M. Fernique, J. P. Conze et J. Gani, Ecole d'Eté de Probabilités de Saint-Flour IV–1974. Edité par P.-L. Hennequin. XI, 293 pages. 1975.

Vol. 481: M. de Guzmán, Differentiation of Integrals in R^n. XII, 226 pages. 1975.

Vol. 482: Fonctions de Plusieurs Variables Complexes II. Séminaire François Norguet 1974–1975. IX, 367 pages. 1975.

Vol. 483: R. D. M. Accola, Riemann Surfaces, Theta Functions, and Abelian Automorphisms Groups. III, 105 pages. 1975.

Vol. 484: Differential Topology and Geometry. Proceedings 1974. Edited by G. P. Joubert, R. P. Moussu, and R. H. Roussarie. IX, 287 pages. 1975.

Vol. 485: J. Diestel, Geometry of Banach Spaces – Selected Topics. XI, 282 pages. 1975.

Vol. 486: S. Stratila and D. Voiculescu, Representations of AF-Algebras and of the Group U (∞). IX, 169 pages. 1975.

Vol. 487: H. M. Reimann und T. Rychener, Funktionen beschränkter mittlerer Oszillation. VI, 141 Seiten. 1975.

Vol. 488: Representations of Algebras, Ottawa 1974. Proceedings 1974. Edited by V. Dlab and P. Gabriel. XII, 378 pages. 1975.

Vol. 489: J. Bair and R. Fourneau, Etude Géométrique des Espaces Vectoriels. Une Introduction. VII, 185 pages. 1975.

Vol. 490: The Geometry of Metric and Linear Spaces. Proceedings 1974. Edited by L. M. Kelly. X, 244 pages. 1975.

Vol. 491: K. A. Broughan, Invariants for Real-Generated Uniform Topological and Algebraic Categories. X, 197 pages. 1975.

Vol. 492: Infinitary Logic: In Memoriam Carol Karp. Edited by D. W. Kueker. VI, 206 pages. 1975.

Vol. 493: F. W. Kamber and P. Tondeur, Foliated Bundles and Characteristic Classes. XIII, 208 pages. 1975.

Vol. 494: A Cornea and G. Licea. Order and Potential Resolvent Families of Kernels. IV, 154 pages. 1975.

Vol. 495: A. Kerber, Representations of Permutation Groups II. V, 175 pages. 1975.

Vol. 496: L. H. Hodgkin and V. P. Snaith, Topics in K-Theory. Two Independent Contributions. III, 294 pages. 1975.

Vol. 497: Analyse Harmonique sur les Groupes de Lie. Proceedings 1973–75. Edité par P. Eymard et al. VI, 710 pages. 1975.

Vol. 498: Model Theory and Algebra. A Memorial Tribute to Abraham Robinson. Edited by D. H. Saracino and V. B. Weispfenning. X, 463 pages. 1975.

Vol. 499: Logic Conference, Kiel 1974. Proceedings. Edited by G. H. Müller, A. Oberschelp, and K. Potthoff. V, 651 pages 1975.

Vol. 500: Proof Theory Symposion, Kiel 1974. Proceedings. Edited by J. Diller and G. H. Müller. VIII, 383 pages. 1975.

Vol. 501: Spline Functions, Karlsruhe 1975. Proceedings. Edited by K. Böhmer, G. Meinardus, and W. Schempp. VI, 421 pages. 1976.

Vol. 502: János Galambos, Representations of Real Numbers by Infinite Series. VI, 146 pages. 1976.

Vol. 503: Applications of Methods of Functional Analysis to Problems in Mechanics. Proceedings 1975. Edited by P. Germain and B. Nayroles. XIX, 531 pages. 1976.

Vol. 504: S. Lang and H. F. Trotter, Frobenius Distributions in GL_2-Extensions. III, 274 pages. 1976.

Vol. 505: Advances in Complex Function Theory. Proceedings 1973/74. Edited by W. E. Kirwan and L. Zalcman. VIII, 203 pages. 1976.

Vol. 506: Numerical Analysis, Dundee 1975. Proceedings. Edited by G. A. Watson. X, 201 pages. 1976.

Vol. 507: M. C. Reed, Abstract Non-Linear Wave Equations. VI, 128 pages. 1976.

Vol. 508: E. Seneta, Regularly Varying Functions. V, 112 pages. 1976.

Vol. 509: D. E. Blair, Contact Manifolds in Riemannian Geometry. VI, 146 pages. 1976.

Vol. 510: V. Poènaru, Singularités C^∞ en Présence de Symétrie. V, 174 pages. 1976.

Vol. 511: Séminaire de Probabilités X. Proceedings 1974/75. Edité par P. A. Meyer. VI, 593 pages. 1976.

Vol. 512: Spaces of Analytic Functions, Kristiansand, Norway 1975. Proceedings. Edited by O. B. Bekken, B. K. Øksendal, and A. Stray. VIII, 204 pages. 1976.

Vol. 513: R. B. Warfield, Jr. Nilpotent Groups. VIII, 115 pages. 1976.

Vol. 514: Séminaire Bourbaki vol. 1974/75. Exposés 453 – 470. IV, 276 pages. 1976.

Vol. 515: Bäcklund Transformations. Nashville, Tennessee 1974. Proceedings. Edited by R. M. Miura. VIII, 295 pages. 1976.

Vol. 516: M. L. Silverstein, Boundary Theory for Symmetric Markov Processes. XVI, 314 pages. 1976.

Vol. 517: S. Glasner, Proximal Flows. VIII, 153 pages. 1976.

Vol. 518: Séminaire de Théorie du Potentiel, Proceedings Paris 1972–1974. Edité par F. Hirsch et G. Mokobodzki. VI, 275 pages. 1976.

Vol. 519: J. Schmets, Espaces de Fonctions Continues. XII, 150 pages. 1976.

Vol. 520: R. H. Farrell, Techniques of Multivariate Calculation. X, 337 pages. 1976.

Vol. 521: G. Cherlin, Model Theoretic Algebra – Selected Topics. IV, 234 pages. 1976.

Vol. 522: C. O. Bloom and N. D. Kazarinoff, Short Wave Radiation Problems in Inhomogeneous Media: Asymptotic Solutions. V. 104 pages. 1976.

Vol. 523: S. A. Albeverio and R. J. Høegh-Krohn, Mathematical Theory of Feynman Path Integrals. IV, 139 pages. 1976.

Vol. 524: Séminaire Pierre Lelong (Analyse) Année 1974/75. Edité par P. Lelong. V, 222 pages. 1976.

Vol. 525: Structural Stability, the Theory of Catastrophes, and Applications in the Sciences. Proceedings 1975. Edited by P. Hilton. VI, 408 pages. 1976.

Vol. 526: Probability in Banach Spaces. Proceedings 1975. Edited by A. Beck. VI, 290 pages. 1976.

Vol. 527: M. Denker, Ch. Grillenberger, and K. Sigmund, Ergodic Theory on Compact Spaces. IV, 360 pages. 1976.

Vol. 528: J. E. Humphreys, Ordinary and Modular Representations of Chevalley Groups. III, 127 pages. 1976.

Vol. 529: J. Grandell, Doubly Stochastic Poisson Processes. X, 234 pages. 1976.

Vol. 530: S. S. Gelbart, Weil's Representation and the Spectrum of the Metaplectic Group. VII, 140 pages. 1976.

Vol. 531: Y.-C. Wong, The Topology of Uniform Convergence on Order-Bounded Sets. VI, 163 pages. 1976.

Vol. 532: Théorie Ergodique. Proceedings 1973/1974. Edité par J.-P. Conze and M. S. Keane. VIII, 227 pages. 1976.

Vol. 533: F. R. Cohen, T. J. Lada, and J. P. May, The Homology of Iterated Loop Spaces. IX, 490 pages. 1976.

Vol. 534: C. Preston, Random Fields. V, 200 pages. 1976.

Vol. 535: Singularités d'Applications Differentiables. Plans-sur-Bex. 1975. Edité par O. Burlet et F. Ronga. V, 253 pages. 1976.

Vol. 536: W. M. Schmidt, Equations over Finite Fields. An Elementary Approach. IX, 267 pages. 1976.

Vol. 537: Set Theory and Hierarchy Theory. Bierutowice, Poland 1975. A Memorial Tribute to Andrzej Mostowski. Edited by W. Marek, M. Srebrny and A. Zarach. XIII, 345 pages. 1976.

Vol. 538: G. Fischer, Complex Analytic Geometry. VII, 201 pages. 1976.

Vol. 539: A. Badrikian, J. F. C. Kingman et J. Kuelbs, Ecole d'Eté de Probabilités de Saint Flour V-1975. Edité par P.-L. Hennequin. IX, 314 pages. 1976.

Vol. 540: Categorical Topology, Proceedings 1975. Edited by E. Binz and H. Herrlich. XV, 719 pages. 1976.

Vol. 541: Measure Theory, Oberwolfach 1975. Proceedings. Edited by A. Bellow and D. Kölzow. XIV, 430 pages. 1976.

Vol. 542: D. A. Edwards and H. M. Hastings, Čech and Steenrod Homotopy Theories with Applications to Geometric Topology. VII, 296 pages. 1976.

Vol. 543: Nonlinear Operators and the Calculus of Variations, Bruxelles 1975. Edited by J. P. Gossez, E. J. Lami Dozo, J. Mawhin, and L. Waelbroeck, VII, 237 pages. 1976.

Vol. 544: Robert P. Langlands, On the Functional Equations Satisfied by Eisenstein Series. VII, 337 pages. 1976.

Vol. 545: Noncommutative Ring Theory. Kent State 1975. Edited by J. H. Cozzens and F. L. Sandomierski. V, 212 pages. 1976.

Vol. 546: K. Mahler, Lectures on Transcendental Numbers. Edited and Completed by B. Diviš and W. J. Le Veque. XXI, 254 pages. 1976.

Vol. 547: A. Mukherjea and N. A. Tserpes, Measures on Topological Semigroups: Convolution Products and Random Walks. V, 197 pages. 1976.

Vol. 548: D. A. Hejhal, The Selberg Trace Formula for $PSL(2,\mathbb{R})$. Volume I. VI, 516 pages. 1976.

Vol. 549: Brauer Groups, Evanston 1975. Proceedings. Edited by D. Zelinsky. V, 187 pages. 1976.

Vol. 550: Proceedings of the Third Japan – USSR Symposium on Probability Theory. Edited by G. Maruyama and J. V. Prokhorov. VI, 722 pages. 1976.

Vol. 551: Algebraic K-Theory, Evanston 1976. Proceedings. Edited by M. R. Stein. XI, 409 pages. 1976.

Vol. 552: C. G. Gibson, K. Wirthmüller, A. A. du Plessis and E. J. N. Looijenga. Topological Stability of Smooth Mappings. V, 155 pages. 1976.

Vol. 553: M. Petrich, Categories of Algebraic Systems. Vector and Projective Spaces, Semigroups, Rings and Lattices. VIII, 217 pages. 1976.

Vol. 554: J. D. H. Smith, Mal'cev Varieties. VIII, 158 pages. 1976.

Vol. 555: M. Ishida, The Genus Fields of Algebraic Number Fields. VII, 116 pages. 1976.

Vol. 556: Approximation Theory. Bonn 1976. Proceedings. Edited by R. Schaback and K. Scherer. VII, 466 pages. 1976.

Vol. 557: W. Iberkleid and T. Petrie, Smooth S^1 Manifolds. III, 163 pages. 1976.

Vol. 558: B. Weisfeiler, On Construction and Identification of Graphs. XIV, 237 pages. 1976.

Vol. 559: J.-P. Caubet, Le Mouvement Brownien Relativiste. IX, 212 pages. 1976.

Vol. 560: Combinatorial Mathematics, IV, Proceedings 1975. Edited by L. R. A. Casse and W. D. Wallis. VII, 249 pages. 1976.

Vol. 561: Function Theoretic Methods for Partial Differential Equations. Darmstadt 1976. Proceedings. Edited by V. E. Meister, N. Weck and W. L. Wendland. XVIII, 520 pages. 1976.

Vol. 562: R. W. Goodman, Nilpotent Lie Groups: Structure and Applications to Analysis. X, 210 pages. 1976.

Vol. 563: Séminaire de Théorie du Potentiel. Paris, No. 2. Proceedings 1975–1976. Edited by F. Hirsch and G. Mokobodzki. VI, 292 pages. 1976.

Vol. 564: Ordinary and Partial Differential Equations, Dundee 1976. Proceedings. Edited by W. N. Everitt and B. D. Sleeman. XVIII, 551 pages. 1976.

Vol. 565: Turbulence and Navier Stokes Equations. Proceedings 1975. Edited by R. Temam. IX, 194 pages. 1976.

Vol. 566: Empirical Distributions and Processes. Oberwolfach 1976. Proceedings. Edited by P. Gaenssler and P. Révész. VII, 146 pages. 1976.

Vol. 567: Séminaire Bourbaki vol. 1975/76. Exposés 471–488. IV, 303 pages. 1977.

Vol. 568: R. E. Gaines and J. L. Mawhin, Coincidence Degree, and Nonlinear Differential Equations. V, 262 pages. 1977.

Vol. 569: Cohomologie Etale SGA $4\frac{1}{2}$. Séminaire de Géométrie Algébrique du Bois-Marie. Edité par P. Deligne. V, 312 pages. 1977.

Vol. 570: Differential Geometrical Methods in Mathematical Physics, Bonn 1975. Proceedings. Edited by K. Bleuler and A. Reetz. VIII, 576 pages. 1977.

Vol. 571: Constructive Theory of Functions of Several Variables, Oberwolfach 1976. Proceedings. Edited by W. Schempp and K. Zeller. VI, 290 pages. 1977

Vol. 572: Sparse Matrix Techniques, Copenhagen 1976. Edited by V. A. Barker. V, 184 pages. 1977.

Vol. 573: Group Theory, Canberra 1975. Proceedings. Edited by R. A. Bryce, J. Cossey and M. F. Newman. VII, 146 pages. 1977.

Vol. 574: J. Moldestad, Computations in Higher Types. IV, 203 pages. 1977.

Vol. 575: K-Theory and Operator Algebras, Athens, Georgia 1975. Edited by B. B. Morrel and I. M. Singer. VI, 191 pages. 1977.

Vol. 576: V. S. Varadarajan, Harmonic Analysis on Real Reductive Groups. VI, 521 pages. 1977.

Vol. 577: J. P. May, E_∞ Ring Spaces and E_∞ Ring Spectra. IV, 268 pages. 1977.

Vol. 578: Analyse Complexe. I. Séminaire Pierre Lelong (Analyse) Année 1975/76, II. Journées sur les Fonctions Analytiques, Toulouse 1976. Edité par P. Lelong. VI, 327 pages. 1977.

Vol. 579: Combinatoire et Représentation du Groupe Symétrique, Strasbourg 1976. Proceedings 1976. Edité par D. Foata. IV, 339 pages. 1977.